An Introduction to Numerical Classification

An Introduction to Numerical Classification

H. T. CLIFFORD

Botany Department
University of Queensland
St. Lucia, Brisbane, Australia

W. STEPHENSON

Zoology Department
University of Queensland
St. Lucia, Brisbane, Australia

Academic Press

New York San Francisco London 1975

A Subsidiary of Harcourt Brace Jovanovich, Publishers

ACADEMIC PRESS, INC.
111 Fifth Avenue, New York, New York 10003

United Kingdom Edition published by
ACADEMIC PRESS, INC. (LONDON) LTD.
24/28 Oval Road, London NW1

Library of Congress Cataloging in Publication Data

Clifford, Harold Trevor.
An introduction to numerical classification.

Bibliography: p.
Includes index.
1. Numerical taxonomy. I. Stephenson, William, Date II. Title.
QH83.C565 574'.01'2 74-17983
ISBN 0-12-176750-7

PRINTED IN THE UNITED STATES OF AMERICA

Contents

Preface ix

1 INTRODUCTION

Text 1

2 CLASSIFICATION BY STRUCTURE

A. Naming 3
B. The Higher Categories: Arbitrary Divisions 4
C. The Evolutionary Framework: "The New Systematics" 7
D. Models Showing Taxonomic Relationships 8
E. The Taxonomic Continuum 10
F. Critique of Classic Taxonomy 12

3 CLASSIFICATION BY PROXIMITY

A. Biogeographical Classification 15
B. Ecological Classification 16

4 GENERAL COMMENTS ON CLASSIFICATION

A. Continua in Nonbiological Situations 25
B. What Classification Involves 26

5 NUMERICAL APPROACHES TO CLASSIFICATION

A. Introduction 31
B. Types of Data 32

6 MEASURES OF SIMILARITY AND DIFFERENCE

A. General 49
B. Coefficients of Similarity 51
C. Coefficients of Association 61
D. Euclidean Distance as a Dissimilarity Measure 65
E. Information Theory Measures of Similarity/Dissimilarity 67
F. Probabilistic Measures 77
G. Further Properties of Similarity Measures 77

7 REDUCTION, TRANSFORMATION, AND STANDARDIZATION OF DATA

A. General 83
B. Data Reduction 85
C. Data Transformation 89
D. Data Standardization 93
E. Reduction, Transformation, and Standardization of Taxonomic Data 97
F. Discussion of Data Manipulation 98

8 SIMILARITY MATRICES AND THEIR ANALYSIS

A. Visual Matrices—Trellis Diagrams 99
B. Classificatory Strategies in General 100
C. Monothetic Divisive Hierarchical Clustering Methods 101
D. Agglomerative Polythetic Hierarchical Clustering Methods 104
E. Nonhierarchical Clustering, Clumping, Graphs, and Minimum Spanning Trees 117

9 THE HANDLING AND INTERPRETATION OF THE RESULTS OF COMPUTER CLASSIFICATIONS

A. General Comparison and Interpretation of Results 125
B. Application of Significance Tests 136
C. Combination of Strategies 138
D. Dendrograms 138

10 DIFFICULTIES IN NUMERICAL CLASSIFICATION

A. Objectives in Classification 143
B. Choice of Data 146
C. Choice of Strategy 147
D. Presentation of Results 148
E. The Time Factor in Ecological Analyses: Multidimensional Data 149
F. Machine Dependency 153

11 RELATIONSHIPS OF SPECIES TO EXTRINSIC FACTORS IN ECOLOGICAL ANALYSES

Text 155

12 DIVERSITY AND CLASSIFICATION

A. Measures Based on Numbers of Species 158
B. Measures Based on the Proportions of Species Present 159
C. Measures of Evenness or Equitability 161
D. Importance of Dominance in Diversity Measures 162
E. Interpretations of Diversity 163
F. Alpha, Beta, and Gamma Diversities 166
G. Habitat Width and Habitat Overlap 167

13 MULTIVARIATE ANALYSIS

A. Introduction 169
B. Principal Component Analysis 170

C. Factor Analysis 178
D. Principal Coordinate Analysis 182
E. Canonical Variate Analyses 183
F. Canonical Correlation Analysis 184
G. Interpreting Ordinations 186

APPENDIX

A. Information Theory Measures of Diversity 191
B. Partitioning of Diversity of the Information Content of a Two-Way Table 196
C. Information Gain with Multistate Attributes 203
D. Information Measures and Interdependence 206

BIBLIOGRAPHY 209

Subject Index 225

Preface

Although this book is written primarily for biologists, we anticipate that its contents may appeal to a wider range of readers. In biology there has been an embarrassment of riches due to voluminous masses of data for longer than in most other sciences and, as a result, biologists have been forced to consider classificatory problems earlier than many other workers. As data accumulates in all sciences, so does the realization that it needs simplification by such techniques as classification, dissection, and ordination. We say later that attempts to find satisfactory breaks in continuous sets of data are part and parcel of human activities. These and related matters, set in a biological context, are the themes of this book.

Until some fifteen years ago, the use of numerical methods in biology generally meant the application of standard statistics and standard mathematics to biological problems. While a variety of numerical methods of examining the similarities and differences between taxa were part and parcel of biomathematics, extensive use of numerical methods was hindered by the sheer labor of the calculations. With the advent of computers it became practical to use a wider range of techniques than had hitherto been possible and to apply them more extensively to taxonomic and ecological data.

Many of these newer methods deviate considerably from standard statistics, and, while all have a mathematical basis, in many cases their mathematical properties are not fully known. For example, in many cases it is not yet possible or not appropriate to apply tests of significance to results, and the general descriptive term "nonprobablistic" has been coined for these situations.

There have been criticisms of the methods from many directions. Pure mathematicians object to their impurity. The objections of taxonomists have concentrated on the undesirability of replacing their intuitive and experienced conclusions with those of a machinelike device.

In dealings with our biological colleagues it has become evident to us that the words "numerical analyses" and "computers" are now charged with emotional significance. Two groups of biologists have become apparent—the "traditionalists" and the "avante-garde." The former condemn computer methodology without knowing too much of what it is all about. We have sympathy with this viewpoint because we were both raised as classic biologists and because we also know how difficult it is for an average biologist to become at all familiar with the methods. This is in part because of a special set of terms which has arisen whose meanings are difficult to determine, and in part because of the difficulties most biologists have with mathematical symbolism. We use the words "mathematical symbolism" rather than "mathematical ideas" because we believe that the basic ideas behind numerical analyses are not difficult to comprehend and because we also believe that most biologists have instinctive powers of mathematical judgment.

We hope that the book also has value to the "avante-garde." In particular, it indicates that a very wide range of techniques is available for handling any one set of data. We hope a sufficient warning is sounded against the adoption of the single method heard of by an inexperienced worker. We also warn strongly against the assumption that because results have come from a computer these must necessarily reveal a final and fundamental truth.

It seems that there is no hard and fast division between the "subjectivity" of classic methods and the "objectivity" of numerical methods because judgments are involved in both cases. The "traditionalists" can thus condemn numerical analyses on the grounds that there is subjective choice of the method which will produce the answer which was wanted. The truth of the matter is that methods are chosen which give the *kind* of answers which are likely to be meaningful but not a precise and specific answer which was "required." The great value of the present methods is that they can handle vastly more data than can the unaided brain; under these circumstances no precise and specific answer can be known in advance.

For many years the present techniques were called "numerical taxonomy," but it is clear from the text that they have been widely applied for many years in ecology. For this reason we have avoided "numerical taxonomy" in the title and also its synonym "taxometry."

We have attempted to give almost equal stress to taxonomic and ecological aspects. We have tried to show some of the interactions between analyses in the two areas and why, for example, the appropriate methods in taxonomy are not necessarily the most appropriate for ecological use. The differing properties of the methodologies are thereby exposed. Ecological problems are more complicated than those in taxonomy possibly because the continuity imposed by successive generations is lacking and so further developments can be expected. While ecological work can be more "difficult," there are compensations; particularly, there is a wider range of methods for determining the relative values of a series of classifications, largely by reference to environmental criteria.

Our general approach has been to simplify the problems of entry of biologists into what, for many, is likely to be a new field. We have both given several courses at tertiary institutions dealing with the matter of the book, and we both know the glazed looks which can arise. To minimize these reactions we attempt to explain the rationale of numerical analyses by means of geometrical models or worked examples, and have avoided where possible extensive algebraic symbolism.

Our personal interests have largely dictated the areas we have covered. The taxonomy has a botanical bias (H.T.C.) and the ecology a bias to marine biology (W.S.), and we realize that zoological taxonomy and terrestrial ecology have not been covered at all intensively. Clearly our text is illustrative rather than exhaustive. This is reflected in the Bibliogrophy, which although extensive, is far from all embracing. For supplementary reading we recommend for taxonomy generally "Numerical Taxonomy. The Principles and Practice of Numerical Classification," by P. H. A. Sneath and R. R. Sokal (1973; Freeman, San Francisco), and for plant ecology "Quantitative Plant Ecology," 2nd ed. by P. Greig-Smith (1964; Butterworth, London).

For our own convenience and that of our readers, our citations are almost exclusively from publications in English. We do realize that an imposing bibliography could be compiled from non-English publications.

It is our pleasure to record our indebtedness to the staff of the C.S.I.R.O. Division of Computing Research, Canberra, with whom both of us have enjoyed cordial relationships and cooperation for a number of years. Without such goodwill neither of us would have progressed far with the subject and this text could not have been written. We are particularly indebted to Dr. Bill (W.T.) Williams for his kindly good humor and patience in explaining to us much of the mathematics involved and for being actively associated with our research programs. We are also grateful to Professor J. D. Burr of the University of New England for many illuminating insights into the mathematics of classificatory procedures.

Finally, we pay tribute to the skill of Kathy Haviland and Jane Hodgkinson for their patience and skill in converting our handwritten text into readable typescript, and to Stephen Cook for preparing the illustrations.

It is to be noted our names are in alphabetical order of surname; there is no senior author.

H.T. Clifford
W. Stephenson

An Introduction to Numerical Classification

1

Introduction

While all sciences are now in a position of embarrassment due to a multiplicity of factual information, the biological sciences have felt embarrassment the longest. This is because so many organisms occur in so many situations and because so many of them have for so long attracted the attention of man. Long before the invention of writing and the coincident beginnings of history, prehistoric man must have amassed an encyclopedic knowledge of his organic environment in terms of what lived where and whether it was friend or foe, food or poison.

The existence of such concepts as "food" implies that organisms have been put into groups and hence implies classification. Simpson (1961) has stated that the necessity for aggregating things into categories is a general characteristic of living organisms and notes that *Amoeba*, like man, cannot exist without appreciation of the category "food." Human thought in general, as reflected in human language, seems greatly dependent on the recognition of groups, particularly the groups of objects which give our collective nouns. We should note that the collective noun "tree" involves two mental processes, first of the similarities which trees possess, and second the differences between "trees" and "non-trees."

The necessity for taxonomic classification has been cogently stated by Savory (1970, p. 2): "We find so many different animals in the world that we cannot treat them separately and even if we wanted to do so the task would be beyond the capacity of the human mind and memory. Classification is forced upon us by the limitations of the brain." In Savory's context and in many other works (for example, Mayr *et al.*, 1953), the concept of biological classification is restricted to the grouping of organisms by their structural attributes into taxa—from phylum down to genus and species.

As Simpson (1961) has noted, there are two kinds of relationships which can be used to give general groupings and these are "association by similarity" and "association by contiguity," the first being generally known as "systematics" or "taxonomy." The accepted definitions of these terms logically include "association by contiguity," for example, Simpson's definition of systematics is "The scientific study of the kinds and diversity of organisms and of any and all relationships among them" (Simpson, 1961, p. 7) and of taxonomy is "The theoretical study of classification, including its bases, principles, procedures and rules" (Simpson, 1961, p. 11). Definitions may attempt to establish meanings, but these are conditioned by common usage, and it is significant that Simpson's own book, "Principles of Animal Taxonomy," is virtually restricted to "association by similarity" and the same applies to other standard and well-known works, for example, Mayr *et al.* "Methods and Principles of Systematic Zoology" (1953).

The present work is concerned in part with the resemblances between biological classifications by "similarity" and those by "contiguity" and in part with their differences. These differences are sufficiently important to merit different names. To avoid creating new terms, or hunting through old literature to resurrect others (pursuits all too favored by biologists), we shall use the terms "classification" and "ecological classification," respectively. If "taxonomy" is the classification of structural data, then but for unfortunate preemption "economy" might have been the classification of proximity data.

Taxonomic and ecological classifications resemble each other in the lengths of their history, which in each case goes back to Aristotle. They differ conspicuously in the progress which has been attained by nonnumerical methods. In taxonomy there is an established and workable system which has found general acceptance, in spite of semantic and other difficulties.

In ecology, on the other hand, there are many real differences in opinion on the criteria and end points of classificatory procedures and the scope for numerical approaches seems enhanced. The two disciplines have several fundamentals in common; first, they attempt to divide what appear to be almost continuous systems into discrete entities. (The justification for regarding them as continuous will be given later.) Second, and this applies to most classificatory systems, they compress complex multidimensional relationships into two-dimensional models. In the next two sections we consider the older approaches to taxonomic and ecological data to document these two points, and the established bases from which numerical analyses may offer assistance. In each case the approach is roughly historical.

2

Classification by Structure

A. NAMING

In normal affairs, when we recognize something and incorporate it into our mental processes, we give it a name—a thing which is nameless tends to be disturbing. For individuals, personal names of one or more words may be used; for sets of individuals possessing certain characters in common group names are used and these are usually nouns such as sheep and trees. Equally in classification name-calling (nomenclature) is an integral part of the process. Simpson (1961, p. 9) has said that systematics almost necessarily requires a vocabulary of names for the taxa, though some have proposed the names be replaced by ordered symbols or numbers (Gould, 1958, 1963).

Bartlett (1940) has noted that a culturally primitive group of men in Papua recognized 137 birds by name compared with 138 species which occurred there. It is clear the Papuans are acute observers of their avian fauna and have good eyes for specific differences and presumably this also applied to many prehistoric groups of men.

The accepted scientific binomial system has a counterpart in vernacular languages; in English binomials we put the particular before the general in specific names, for example, "brown rats," just as with given names and surnames. However in other languages, for example, Malayan (Corner, 1952) the general precedes the particular as may be seen from the classification of *Citrus* species given in Table 2.1. With this vernacular use of binomials it is not surprising they were sometimes used in the form, which is now familiar, before the time of Linnaeus. Stearn (1959) says that the binomial names which Linnaeus adopted occur in the works of his predecessors notably Gesner (1516–1609), Caspar Bauhin (1560–1624), Willughby (1635–1673), and Ray (1628–1705). Some of Linnaeus's binomials, for example, *Allium ursinum* (ramsons) and *Lilium candidum* (madonna lily), had even been vernacular names among the Romans some two-

TABLE 2.1

THE SCIENTIFIC, MALAYAN, AND ENGLISH NAMES FOR A GROUP OF CITRUS PLANTS[a]

Scientific name	Malayan name	English name
Citrus aurantifolia	Limau Kapas	Lime
C. grandis	Limau Besar	Pumelo, Shaddock
C. hystrix	Limau Purut	—
C. macroptera	Limau Hantu	False Shaddock
C. medica	Limau Susu	Citron
C. microcarpa	Limau Kesturi	Musk Lime
C. suhuiersis	Limau Manis	—
C. swinglei	Limau Pegar	—

[a] Corner, 1952.

thousand years before. It is generally agreed that Ray and Tournefort (1656–1708) were responsible for the terms "genus" and "species" acquiring their present applications in biology and that Linnaeus accepted and developed their usage.

Linnaeus used binomials, which remain in current as scientific names, and polynomials, which were several-word descriptive and diagnostic names and have been discarded by amplification. Linnaeus regarded the polynomials as the true specific names and Stearn (1959) stated that Linnaeus' binomial nomenclature was almost an accident, in his task of providing definitions and means of identifying genera and species.

B. THE HIGHER CATEGORIES: ARBITRARY DIVISIONS

The early and persistent use of the equivalent of generic and specific names suggests that the naming of systematic categories began at the bottom, by fusion based on similarities—in modern parlance by agglomeration. The complement to agglomeration is division or fission from above on the basis of differences—in modern terms this is divisive classification.

Whether the higher categories came from above or below is less important than that they came, and Aristotle (384–322 B.C.) was apparently the first important contributor. He gave to the animal kingdom the first written formalizations of categories and referred to such major groups as birds, fishes, whales, and insects. The main development of these categories was by Linnaeus, following such earlier workers as Ray. The elaborations of Linnaeus, a system from phylum and class down to genus and species, are well known and almost universally adopted in the animal kingdom. With plants no such clear-cut divisions have been accepted and although groups now accepted as families were recognized as long ago as the time of Theophrastus (380–320 B.C.), the subdivision or aggregation of the plant kingdom has as yet produced no generally accepted set of phyta.

There have been many disagreements on which levels in the hierarchy are "natural" and which are "artificial." To most taxonomists and practically all ecologists, the species is the vital taxon. Most laymen accept that it is a "real" thing and can recognize most of the animals in the surroundings as belonging to "real" species. Simpson (1961, p. 115) has outlined the situation as it applies to the living fauna in these words: ". . . the species (genetically defined) is the one taxon that is usually nonarbitrary both as to inclusion and as to exclusion."

In the definitions of taxonomic levels there is usually an arbitrary element. For example, Mayr *et al.* (1953, p. 51) define a family: ". . . as a systematic category including one genus or a group of genera of common phylogenetic origin, which is separated from other families by a decided gap." Avoiding the generally unproved assumptions about a common phylogenetic origin, the critical words are "a decided gap," and these also appear in the above authors' definition of a genus. In practice, a definition in similar terms is the only one for a species which works in the majority of cases. Cain (1953, p. 82), for example, in noting that there are four main practical meanings of "species," lists as his first: "A group of specimens which is considered sufficiently different morphologically from the most closely related forms known, to be given a specific name." On bases such as these, family, generic, and specific boundaries depend on how one defines "a decided gap" or "sufficiently different morphologically." The criteria differ from one group of organisms to another and being incapable of comprehensive definition are left to the good sense, intuition, and experience of a taxonomist familiar with the group, who defines them within a group context. It is evident that there is an arbitrary flavor throughout and every group specialist will know that while species tend to be "fixed" and "good," other levels have shown flexibility.

This is due mostly to increased knowledge, both of the numbers of organisms and of the characters on which species may be recognized. Stearn (1959, p. 8) has stated: "Modern estimates put the number of known species of flowering plants as between 250,000 and 380,000 many times more than he [Linnaeus] thought possible for the whole vegetable kingdom . . . . between 1900 and 1955 botantists described some 198,000 species of flowering plants as new, undoubtedly with undue optimism! The number of living species of Insecta is estimated at about 754,000–850,000 and of animals as a whole 930,000–1,120,000." Stearn also noted that it is fortunate for biology that Linnaeus passed his life in blissful ignorance of such frightening figures and the point is still valid even if the figures have been overstated. Mayr *et al.* (1953) note that Silvestri estimated that there were 3 million possible insect species, each with an average of five distinct phases or stages requiring description, thus requiring 15 million descriptions for the insects alone. The details of descriptions have increased in a similar way, and instead of the polynomials of Linnaeus we have the several pages of modern works. With the wisdom of hindsight we realize these sometimes err by overbrevity.

It is interesting to note that with respect to the flowering plants the approach of Linnaeus was almost completely numerical. Species were allotted to groups on the basis of the numbers of anthers and carpels per flower. The anther number was given first and group names such as Hexandria trigyna, which included the lilies and their allies among others, originated. The groups that resulted from this approach were often "unnatural" assemblages of plants and were rarely acceptable to plant taxonomists.

Linnaeus' system as applied to the animal kingdom has had to be modified to cope with the increased number of species, and the greatest changes have been in the middle ranks of the hierarchy. Most of Linnaeus' genera have become families, a category he did not recognize and one still rapidly increasing in numbers at the beginning of the twentieth century. At the end of the nineteenth century, there were about 1700 families of all animals, but by 1932 there were 100 recognized families of insects alone (Mayr *et al.*, 1953, p. 52).

Although greatly modified, the Linnaean system has not broken down. While it may be almost impossible to define the terms used ("species," "family," etc.) and while these may be difficult for the nonexpert to distinguish, in most cases they appeal to our inherent classificatory sensibilities. Simpson (1961, p. 57) has stated: "The feelings and agreements of taxonomists have impelling pragmatic significance but no direct theoretical value. In fact much of the . . . discussion . . . has . . . been an attempt to find some theoretical basis for these personal and subjective results. . . . Part of the basis, although not all of it, is the simple fact that readily recognisable and definable groups of associated organisms do really occur in nature."

C. THE EVOLUTIONARY FRAMEWORK: "THE NEW SYSTEMATICS"

The contribution of Darwin to structural classification is sufficiently well known not to require elaboration. It provided a conceptual framework to parallel the hierarchies of the systematists, and thus stimulated their efforts. As Mayr *et al.* (1953, p. 9) have said: "This was an exciting period in the history of taxonomy, not only were new species and genera discovered daily, but with reasonable frequency even new families or orders. The reward of such exciting discoveries attracted the keenest minds to the field of taxonomy." The wealth of new major discoveries was largely exhausted before the end of the nineteenth century, and Mayr *et al.* (1953, p. 9) continue: "It appears, in retrospect, as the most retrogressive period in the history of taxonomy . . .. Part of the disrepute into which taxonomy fell during the latter part of the nineteenth and early twentieth century was caused by the activities of those who unnecessarily split well-known and well-founded taxonomic categories, thereby hopelessly concealing natural affinities." We feel that this statement shows some personal bias. Because taxonomic categories are well known and well founded it does not mean that with further knowledge, they should remain inviolate.

That taxonomy did fall into disrepute is undoubted. Savory (1970) claims one cause to be the formal, almost lifeless mode of treatment in zoology texts, but instead of being a cause, this could have been a consequence; it might well appear that taxonomic zoologists had flogged to death the hobbyhorse of phylogeny.

In the 1930's interest revived and was stimulated in the early 1940's by the two books of J.S. Huxley, "The New Systematics" (1940) and "Evolution the Modern Synthesis" (1942). The main thesis was that populations rather than individuals were the units of systematics, and that species should not be tied to the type specimen (or specimens) of the original author. Variability within species became studied more intensively, and statistical methods of distinguishing between species and within species were urged. Much of this approach was not really new, and the desirability of preserving large series of specimens of each species was well understood by museum curators. As long ago as 1854, Baird (quoted by Mayr *et al.* 1953, p. 11) stated, in relation to the Smithsonian Institution: "As the object of the Institution in making its collections is not merely to possess the different species, but also to determine their geographical distribution, it becomes important to have as full series as practicable from each locality . . . ."

The "New Systematics" was valuable in focusing attention on the mechanisms of speciation and linked the activities of geneticists and biogeographers to the mainstreams of the existing biology. Nevertheless taxonomy failed to regain its earlier vigor, and the impression created by the "New Systematics" was less than its proponents may have expected. Possibly the "average biologist" was frightened by the statistics, although those normally considered can with caution be used simply and without full understanding of their derivation. As listed in Mayr *et al.*, they are at the level of dissection of a bimodal curve, histograms, scatter diagrams, standard deviation, chi-square, and coefficients of difference.

Part of the relative lack of interest in the "New Systematics" may be because of a changing emphasis in the rationale for identifying species. Fewer people seem interested in the exhaustive exploration of the affinities of taxa, and more regard identification as a means to another end, for example, in ecology. There is no doubt that such studies depend entirely on taxonomic expertise. Elton (1947), quoted in Mayr *et al.* (1953, p. 20), says: "[Accurate identification] is the essential basis of the whole thing: without it the ecologist is helpless, and the whole of his work may be rendered useless." Chace (1969) rightly refers to ". . . a major and virtually insuperable problem which plagues any ecological survey in tropical waters today; that is, the imperfect knowledge of the existing fauna . . .." If the taxonomist has become the servant of the ecologist, he need not feel any need to apologize, but quite the reverse. In this era of fervor for preservation of environments, of species, and of diversity, it is amazing that the only individuals who can tell us what species occupy which environments are a dwindling band of semimaligned but dedicated taxonomists.

D. MODELS SHOWING TAXONOMIC RELATIONSHIPS

Visual representations of taxonomic relationships (apart from written visual presentation) seem to have begun in a single dimension with a single line divided into segments (Bonnet, 1745, quoted by Simpson, 1961). While this has obvious shortcomings it should be noted that the input of human thoughts and of computer data occurs similarly in a linear fashion.

Two-dimensional visualization seems to have begun with Buffon (1707–1788), but the earliest published tree diagram was probably that of Lamarck (1809), and a diagrammatic tree is the only figure in "The Origin of Species" (Darwin, 1859). Most probably trees proliferated as a consequence of Darwin's work which stressed interrelationships of groups by descent and had an obvious parallel with interrelationships of human beings by genealogy—a tree is merely an upsidedown genealogy.* Haeckel (1866) elaborated and exploited the tree visualization in a strictly two-dimensional way, and Lam (1936) should be consulted for the "phylogeny of the phylogenetic tree."

Three-dimensional phylogenetic diagrams were discussed by Lam, and an elegant example from Milne and Milne (1939) of the use of perspective is given by Mayr *et al.* (1953, p. 176). These diagrams allow the visualization of relationships which are obscured in a two-dimensional presentation, for example, those of parallel and converging evolution.

Meanwhile, Mayr *et al.* (1953, pp. 575-578, and 175) introduced a term in frequent current use—"dendrogram." This they define as ". . . a diagrammatic illustration of relationship based on degree of similarity (morphological or otherwise) . . .." Via an example (Fig. 39 in Mayr *et al.*, 1953), they show that angles between branches are of no importance, but points of origin of branches are very important. It is clear that in coining this term the authors were reacting against the creation of phylogenetic trees based on a lack of paleontological evidence. It has also become clear that the term "dendrogram" should be restricted to branching two-dimensional models; this is necessary to distinguish them from ordination diagrams.

A further perceptual model, that of *taxonomic distance*, is vital to an understanding of numerical methods, but preceded their main development, going back at least as far as Heincke (1898). This concept is explained by Sokal and Sneath (1963) and may be summarized as follows. Suppose two species differ by a single character, this can be represented symbolically by two points separated by a line one unit long. The length of this line is the taxonomic distance. For a second character (attribute) we can either add another unit to the existing line or draw another line at right angles to the first—it is the latter course we follow. We now have the two species separated by the hypotenuse of a right-angled triangle whose other sides are each 1 unit long—the taxonomic distance between them is now $\sqrt{2}$ (see Fig. 2.1). For three characters three axes are necessary; our new right-angled triangle will have one side $\sqrt{2}$, one side unity, and the third (the hypotenuse) which is the taxonomic distance separating the taxa will be $\sqrt{3}$. For m characters we need a model in m-dimensional space and the maximum taxonomic distance possible between two species is $\sqrt{m}$. The concept of taxonomic distance has had its main value in numerical analysis, but Westoll (1956) has used it in an illuminating insight into problems in paleontology.

* It should be noted that family trees are two-dimensional only as a result of two conventions. The first is that offspring have only two attributes, maleness and femaleness, and thus other divergences in genetic make-up can be ignored. The second is that "ancestor-less" spouses can be inserted into the tree. When a family involves ancestors from other families the situation becomes multidimensional.

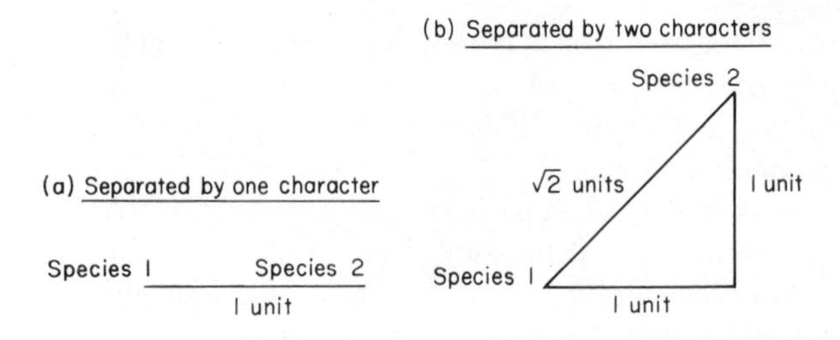

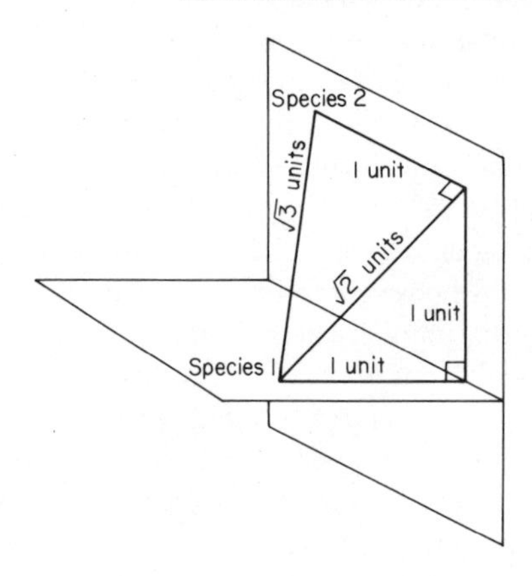

Fig. 2.1. The taxonomic distance between two species differing by 1, 2, or 3 attributes each of one unit.

E. THE TAXONOMIC CONTINUUM

While the living species are "real" entities, when we consider evolutionary history their discreteness is lost. The whole point of phylogeny is that there is a time factor; and if two taxa arise from common ancestry at an earlier period they will have been indistinguishable. Unless developments on the time scale occur in explosive spurts separated by long periods of quiescence, consideration of phylogeny should force acceptance that in classifying by structure we are forcing a discontinuous framework upon a continuous system. This is not the place to review the evidence for explosive spurts in evolution, we can merely say that most of the evidence of evolution at the present time is of a slow and virtually continuous process.

The concept of a phylogenetic continuum has a long history. Simpson (1961, p. 51) states:

> "Lamarck (1744–1829) was, however, the first to maintain clearly and consistently that all taxa have arisen by evolution and are a phylogenetic continuum. His idea of the continuum was far more literal and extensive than that of any modern phylogeneticist. He held in theory that there are no gaps in nature, even between different phylogenetic lineages, but only one continuous progression through which all organisms have tended to pass throughout the history of life and are still doing so today. Taxa, then are merely arbitrary (not therefore necessarily unnatural) subdivisions of the continuum."

Most taxonomists have dealt with the present-day biota in which time is essentially static, but the paleontologists have been alert to the above problems. "The New Palaeontology" published in 1956 and edited by Sylvester-Bradley contains numerous articles in which the problem is ventilated. Sylvester-Bradley himself (1956, p. 2) states: "There remains a third concept that is perhaps more fundamentally different, and this is the result of projecting the biospecies into a third dimension, that of time. It is the concept which Professor George (1956) refers to as the 'chronospecies', and which others have called the 'palaeospecies' (e.g., Cain 1954)." Thomas (1956) states: "However, the neontological concept is inadequate in being virtually static in time, and . . . the palaeontological concept, with the time factor as its prevading life-blood, is the more comprehensive and all-embracing concept." Haldane (1956) states: "Thus in a complete palaeontology all taxonomic distinctions would be as arbitrary as the divisions of a road by milestones. . . . To sum up, the concept of a species is a concession to our linguistic habits and neurological mechanisms (Spurway, 1955)." Ager (1956) continues: "The problem therefore lies in defining the indefinable, and can only be clarified by agreed arbitrary solutions."

Fortunately for the paleontologist the arbitrary solution is usually provided by the nature of his material. Rhodes (1956) states: "In practice, however, as Mayr (1949), Dunbar (1950) and others have pointed out, the existence of an uninterrupted fossil sequence is a rarity, and many, perhaps the great majority, of faunal sequences are divided by 'gaps' of probable depositional hiatus (e.g., Brinckmann, 1929; Stenzel, 1949). In such cases fossil populations have an apparent discontinuity in time, and often also in space."

In summary a system which is basically a continuum becomes discontinuous in its present time stratum, and this permits the neontological taxonomist to recognize his taxa. It is also discontinuous in its sequential time data, which allows the paleontological taxonomist to operate, albeit, more consciously aware of the arbitrary nature of his discontinuities.

F. CRITIQUE OF CLASSIC TAXONOMY

The development of numerical methods, apart from their intrinsic interest, has given insight into principles lying behind classic taxonomy which had not been widely appreciated and indicates some of their shortcomings. Readers interested in extensive criticisms of classic methods by numerical taxonomists are referred to Sokal and Sneath (1963) and Sneath and Sokal (1973).

In normal taxonomy, working from lower levels to higher, we look for highest common factors which are present in one group and absent in another. Combinations of a small number of attributes are usually employed in describing the taxa, and all are chosen with diagnostic importance in mind. (To describe as specific features things which will not distinguish species is usually regarded as wasted effort.) Later if new material is discovered which differs in further and as yet undescribed features from known species, specific descriptions become fuller and further distinctions are made.

Descriptions of taxa are based on fidelity and constancy of characters (attributes, features). Complete fidelity means that a character is shown by members of the particular taxon and of that taxon only, but it is not essential that all members have that particular feature. Complete constancy implies that all the members of a taxon possess the feature; complete constancy is not mandatory, it suffices that a majority of attributes are possessed by each member. Ideally the features used should not only be faithful (i.e., of high fidelity) and constant, but also external and obvious. This is important in saving time and leaving material relatively undamaged. Often in practice the features used at lower levels in the classificatory hierarchy are more obvious than those at upper levels.

Once a taxon has been established by groupings of its highest common factors, remaining attributes can only be used at a subordinate level. This means that bias or weighting is automatically given to the attributes used highest in the hierarchy. Sometimes there are resemblances between species at the lower levels which appear at the opposite sides of a primary dichotomy, but these resemblances appear largely irrelevant. Once a code has been established it becomes difficult to alter as more data are assembled, because typically these data are "low level." To reestablish a new code means relooking at the total mass of data, to reorder the weighting of attributes. To the unaided brain this is a difficult task, much more so than the original codification which involved less data to analyze. For these reasons it could be argued that there is undue conservativism in undertaking taxonomic revisions. Also for these reasons it appears that numerical methods should appeal especially to those undertaking these revisions.

The criticisms of conservativism should be viewed in another light. The taxonomic system can be regarded as a form of filing system, a data-storage and data-recovery device. If the system is changed too frequently, the users become confused, and so good reasons for changes are necessary. One of the criticisms of taxonomists by nontaxonomists is that the system is frequently altered by applications of the laws of priority to nomenclature.

A distinction must now be made between the two basically different approaches to classification. These are the monothetic divisive and the polythetic agglomerative classificatory systems. The first involves subdivision of the entities to be classified by one attribute after another considered in sequence and is the classic method of expression of taxonomy.

The second aggregates individuals into groups on the basis of their overall similarity with respect to several attributes considered simultaneously. Since a polythetic approach involves more than one attribute and since this is usually desired, it is preferable to a monothetic approach to classification. At the present time if the approach is to be polythetic it is also agglomerative. When effective polythetic divisive methods have been developed, this third possibility might be the most satisfactory of the three.

In practice taxonomic systems have usually evolved from the application of both monothetic division and polythetic agglomerative techniques to the problem in hand. Thus when considering a sample of organisms from a taxonomic viewpoint, they may first be sorted into groups such that for each group the members are more similar to one another than they are to the members of other groups. Subsequently the groups might be classified separately in terms of the single attributes which serve to distinguish them.

While in general the human brain is unable to manipulate any considerable mass of data in a polythetic fashion, or if it does so, there is considerable fatigue with an attendant loss of efficiency, such considerations do not apply with the same force to the computer. The computer is no more efficient than its program and so may not be as efficient as a highly trained taxonomist. Nonetheless it does not tire or lose its efficiency and thus has an important role to play in taxonomic studies involving large amounts of data. In these circumstances there is usually a choice of monothetic and polythetic methods and in general, the latter are preferable.

Possibly because this section is titled as if to criticize classic methods in taxonomy, we should conclude by brief mention of the main reverse criticism. Classic workers have criticized numerical methods because they believe that some attributes are more important than others and object that all should be equally weighted. This criticism appears entirely valid—no zoologist would accept that presence/absence of a mandibular palp should be equated with the presence/absence of a chitinous exoskeleton. One of the great disservices that early exponents of numerical taxonomy did to their art was to insist that all attributes should be equally weighted. The reasons for such insistence are diverse but stem principally from the difficulty of deciding on the weight to be assigned to given characters. One solution to this problem is to state the basis of weighting and then the reader may decide for himself as to its merits. For example, it has been argued by Clifford and Goodall (1967) that objective weighting based on the commonness and rarity of attributes is in accord with the spirit though not the letter of the "Adansonian law" of nonweighting adapted so widely by numerical taxonomists. Considerable literature now exists on the different methods of weighting data in numerical taxonomy (see Sneath and Sokal, 1973), and the criticisms outlined above no longer apply with equal force.

3

Classification by Proximity

A. BIOGEOGRAPHICAL CLASSIFICATION

Classification by proximity applies at two levels, large-scale studies are essentially those of the biogeographer involving faunistic and floristic comparisons of large areas, while on a smaller scale they involve ecological classification.

In the former, presence/absence data have been largely used, and this is the approach of the classic biogeographers embodied in such texts as Ekman (1953), Hesse *et al.* (1937), and Darlington (1957). It is also the approach in the great body of faunistic-cum-taxonomic literature dealing with a restricted taxonomic group from a wide area, for example, the portunid crabs of the Indo-West-Pacific (see Stephenson, 1972, and references therein), or with a wider group from a smaller area, for example, the sublittoral crabs of Moreton Bay (Campbell and Stephenson, 1970).

There have been three developments toward quantification. The first is exemplified by the marine ecologists working on rocky shores who have argued that there is a more meaningful biogeographical boundary where a series of dominant organisms is replaced by another than one based on the extreme limits of each species. This approach has been well followed in Australasian work, and on this principle the marine intertidal provinces of the rocky shore biota have been established (see review by Knox, 1963).

The second development has occurred mostly through the efforts of faunistic and taxonomic workers, who have made comparisons between the numbers or proportionality of species common to different areas. Examples in the marine field familiar to us are those of Griffin and Yaldwyn (1968) and of Campbell and Stephenson (1970) and typically involve percentages, for example, the percentage of tropical crabs of Australia known to be widespread throughout the Indo-West-Pacific. In essence these are coefficients of similarity converted to percentages. As in other aspects of biological science, a wide variety of similarity coefficients has developed independently. These were listed in a biogeographical context by Simpson (1960) and their identity with others was later discussed by Cheetham and Hazel (1969). The properties of several of these coefficients are discussed later.

Animal biogeography appears to have developed a numerical flavor much later than in other roughly comparable branches of biological science. For example, it would appear that the advantages of using dissimilarity rather than similarity measures were not recognized until Huheey (1966) used such a measure to express the faunistic difference between two areas.

The third development toward quantification has been the application of numerical classificatory techniques. In animal biogeography this appears to have been initiated by Hagmeir and Stults (1964). The techniques employed were simple and similar approaches to the problem have been presented by Hagmeir (1966) and Peters (1971). More sophisticated studies have been published by Kikkawa and Pearse (1969) working on Australian land birds and Holloway and Jardine (1968) working on Lepidoptera. The former introduced to animal biogeography a series of elegant analyses based on information theory while the latter introduced the concept of "inverse" as well as "normal" analyses. That is, they classified sites in terms of species as well as species in terms of sites.

It is clear that there is opposition to the use of numerical methods by biogeographers. Peters (1971) claims to have used "numerical taxonomic methods" and found them inadequate and unsatisfactory. It is evident from his comments on the work of Kikkawa and Pearse (1969) that some of this reflects his inadequate knowledge of the methods. Sneath and Sokal (1973) should be consulted for further biogeographical studies using numerical analyses.

B. ECOLOGICAL CLASSIFICATION

1. Diffuse Nature of Ecology

Ecological classification is much less well developed than the taxonomy of organisms and this is partly due to the diversity of interests of ecologists. Workers in many fields—for example, population ecology and physiological ecology—either need no classificatory scheme or merely a simple one. It is only in the realms of ecosystems, communities, associations, etc., that the need becomes at all pressing. Even here there are so many different concepts and approaches (see Stephenson, 1973b) that few have felt the need for an all-embracing system which operates in depth. Also it may be significant that the systems which have attempted a formalized exposition to parallel the taxonomic situation—for example, the Braun-Blanquet school with its Pérès-Picard marine counterpart (see Pérès and Picard, 1958)—have not found general acceptance. This is almost certainly due to the main difficulty about ecological data, i.e., they do not lend themselves to easy classification. This applies even when the time factor is excluded, and many ecologists have framed their definitions and determined their sampling with just this exclusion. Every ecologist knows the importance of diurnal, seasonal, and longer term changes and clearly if it is possible such changes should enter the scheme. If they do they create a situation which would be the equivalent of a taxonomic nightmare, with every single taxonomic character varying like the outside of an *Amoeba*. Numerical solutions to the equivalent ecological nightmare of reality are quite recent and are still developing (Williams and Stephenson, 1973; Stephenson *et al.*, 1974).

2. The Use of Names

Before classifying ecological entities into their different types, it is desirable that we should know what it is we are trying to classify. We can simplify by eliminating the "ecosystem" because the "community" is its biological component, but immediately encounter a morass of definitions of "community." Cragg (1953) suggests its meaning is limitless, but Macfadyen (1963) who gives an excellent summary of definitions to that time recognizes three main and some subsidiary definitions. Definitions have ranged from "ecological units of every degree" to organismal, quasi-organismal, and super-organismal entities. Among the variety of concepts which are involved, two are of special importance; these are the biocoenosis concept originated by Möbius (1877) and the biotope concept of Dahl (1908). The former now implies that the discontinuities in the distributional patterns of the biota divide the habitat, while the latter says that the ecological complex is divided by discontinuities in the habitat. [It should be noted that the idea of a biocoenosis has changed. In Möbius terms it was a community whose total of species is mutually linked and selected under the influence of the average conditions of life. Hesse (1924) has virtually synonymized it with the biotope concept by calling it the grouping together of the living beings which dwell in a uniform part of the habitat.]

In an ideal world the biotope and biocoenosis concepts do not conflict, and at the "coarse" levels we can divide the biota into marine, freshwater, and terrestrial by either criterion. When one attempts to delineate actual communities, however, it is clear from the literature that biotope and biocoenosis concepts conflict and entwine in a perplexing manner. Up to Macfadyen's (1963) time it appears that biocoenoses were generally in the ascendancy. For example, of the standard texts, Allee *et al.* (1949, p. 436) say: "... the major community may be defined as a natural assemblage of organisms which, together with its habitat ...." Macfadyen's (1963, p. 185) definition is "... a relatively isolated and discrete population system ... such systems show the properties of constancy of species, dynamic stability and organismic unity," and provided the isolation is not in abiotic terms this is strictly biocoenotic. Fager's (1963, p. 415) two definitions are also strongly biocoenotic, his first being "recurrent organized systems of organisms with similar structure in terms of species presence and abundances ... they are open systems," and his second which he tends to support is "randomly assembled collections of organisms whose ecological tolerances allow them to exist in a particular environment"; this he qualifies by accepting that there is an ecological continuum and that groupings are at best artificial.

Many recent definitions have a biotopic flavor, possibly due to current preoccupancy with ecosystems. In these, where there is an abiotic and a biotic component, there is justification for a classification in abiotic terms. From such ecosystem classifications an abiotic classification of the biota is a logical extension. A recent definition by Whittaker (1970) includes "... an assemblage of populations of plants, animals, bacteria and fungi that live in an environment and interact with one another .... " Here we can delineate the community either by the assemblage or the environment, so it classifies as roughly neutral. However, Odum's (1971, p. 140) definition is solidly biotopic and begins: "A biotic community is any assemblage of populations living in a prescribed area or physical habitat ..." and later includes "... a certain probability that species will occur together."

Of the alternatives of biotope and biocoenosis classification the biotope concept is easier to operate in practice, because it is essentially monothetic, and it still appears to be the one most frequently used. Examples familiar to one of us (W.S.) are afforded by the estuarine complex, in which three main monothetic systems are commonly used. The first is geomorphological and Odum (1971, pp. 352, and 354) has extended Pritchard's (1967) classification to give five types: drowned river valleys, fjord-type, bar-built, from tectonic processes, and river delta estuaries. The second classification is based on mixation of water and gives highly stratified, partially mixed, and completely mixed water estuaries. The third classification is by salinity; and "The Symposium on the Classification of Brackish Waters" (Anonymous, 1959) produced the Venice system which has found acceptance (see, for example, Carriker, 1967; Boesch, 1971). Estuarine waters are classified as euhaline (30–40‰), polyhaline (18–30‰), mesohaline (5–18‰), and oligohaline (0.5–5‰). It is true that these regions reflect to some degree the faunistic divisions of an estuary, but they have all the rigidity of an established system and there may be a tendency to make data conform. They illustrate another problem in ecology generally, and marine ecology in particular. This is the influence of those trained in the "exact sciences" and who enter the marine situation as physical and chemical oceanographers. To an "exact" scientist a classification based on salinity determinations is apparently much preferable to one based on faunistic composition.

However, classifications based on the abiotic environment encounter considerable difficulties when there are no demonstrable discontinuities between habitats, and when there are "multistate attributes." In a marine context this arises over the nature of the bottom deposits, which may merge from mud to sand and which involve many grades of particle sizes. For this reason the studies of benthic faunas have long had a biocoenotic history, beginning with Petersen's (1914) classic work on bottom communities.

Another set of problems arise over the degree to which interactions between organisms are a necessary part of the community concept. Most modern definitions incorporate this, but fortunately this does not obtrude into the classificatory picture. We can agree with Mills (1969) that: "The botanical concept of the association comes closest to the community concept of the zoologist." The two terms have been used as synonyms (Stephenson *et al.*, 1972), as preferable to replacing the term "bottom community" which is so entrenched in the marine literature. In a similar way, in the present context we can virtually synonymize such terms as "formation," "biome," and "major life zone," as used by European ecologists, with the term "association."

Finally we came down to what it is we hope to classify. In the view we have presented, it is associations or communities in terms of their biotic components, in other words biocoenotic classification. There are, as we shall see later, two interacting types of biocoenotic classifications. The first, the "normal," classification is to group sites, not by their abiotic factors (this is the biotopic approach) but by their biota; the species will be the attributes of the sites. The second, the "inverse," classification (see Chapter 5, Section B,1) is to group species by the sites of their occurrence; the sites will be the attributes of the species.

Because of the confusion of terms, a case can be made for paralleling the O. T. U.'s (operational taxonomic units) which are widely used by numerical taxonomists. Presumably, these would be O. E. U. 's (operational ecological units) but they would come in four kinds: O.E.U.'s–SIUSP, viz., sites using species; O.E.U.'s–SPUSI, viz., species using sites; O.E.U.'s–SIUAB, viz., sites using abiotic data; and O.E.U.'s–ABUSI, viz., abiotic data using sites. Such an extended use of initials is unwieldy, and we do not propose to follow it.

To return to generalities, the things we hope to classify will be called *entities*, or sometimes individuals. In ecology the entities will be the samples unless otherwise stated, and their attributes will be the species-in-samples. In taxonomy on the other hand, the entities will be species (or some other appropriate taxonomic level) and the attributes will be the taxonomic characters or features.

In what follows we trace the history of ecological classification, but meanwhile it is evident that there are considerable disagreements about concepts, general agreements that associations are ill-defined, and nothing in the way of accepted names for or accepted levels in classificatory hierarchies.

3. History of Biocoenotic Classification

We divide the history into the following sections: the development as ecology was emerging as a discipline, taking us to the early 1900's; and the later developments of this century.

a. Developments during the Emergence of Ecology

Aristotle (384–322 B.C.) produced the earliest ecological classification of animals and this was divisive and monothetic. His prime division was into water and land animals. The former he divided into: (i) entirely aquatic, (ii) animals that live and feed in water but breathe air and bring forth their young on land; (iii) sea dwellers; (iv) river dwellers; (v) lake dwellers; and (vi) marsh dwellers. This is a classification and can be regarded either as biotopic (which the headings suggest) or biocoenotic; at this level the two coincide. His later divisions of animals indicate a multiplicity of possible alternatives, and in addition to some we would regard as doubtfully ecological, there are contrasts between resident and seasonal birds, gregarious and solitary animals, those with a fixed home or those which are nomadic, those with different diets (carnivorous, graminivorous, or special), nocturnal versus diurnal, those in fields or mountains or frequenting the abodes of men, and so on. An interesting couplet, quoted at third hand is "Some are salacious, e.g. the cock; others are inclined to chastity." It is evident that Aristotle was well aware of the variety of niche alternatives, that he recognized the trophic differences which are such a large part of current ecological thinking, and that he realized the importance of the time scale. His exposition is a good illustration of the diffuse nature of the subject.

Ramaley (1940) considers Theophrastus as the first ecologist and says that he wrote of the communities in which plants are associated and of the relations of plants to each other and to their abiotic environments. Greene (1909) states that Theophrastus recognized the natural associations of plants and related them to their habitats, for example, marine aquatics, marine littoral, plants of deep fresh water, those of shallow lake shores, those of wet stream banks, and those of marshes. While Theophrastus recognized associations of plant species it is uncertain whether he divided the environment by virtue of its vegetation, or "fitted" the vegetation to an environmental classification.

Pliny the Elder (A.D. 23–79) in his "Natural History" had a broad environmental division and recognized the categories of terrestrial, aquatic, and flying animals, but it appears that matters then remained dormant until the sixteenth century when ecological data was recorded in botanical works such as Cordus (1515–1544) as outlined by Greene (1909). Nordenskiold (1928) has claimed that Linnaeus (1707–1778) originated all of what is now called ecological zoology and botany, but Lankester (1889) gives greater prominence to Buffon (1707–1778). Gradual accumulation of habitat data was a natural consequence of the recognition of new species following Linnaeus' time, but it is doubtful if this was more than an accumulation.

While not immediately related to classification, the introduction of statistical approaches to ecology is important. It began much earlier in ecology than in taxonomy. While Malthus (1798) is often regarded as the starting point of population ecology, Strangeland (1904) quotes Machiavelli of 275 years earlier, Botero in 1590, Hale in 1677, and others as employing statistical concepts. Nevertheless there was a long interval before mathematical concepts were introduced into synecological classification—the earliest publication known to us is Kulczynski (1928).

The earliest use of the term "associations" appears to have been by von Humboldt (1805) and then Braun-Blanquet (1932) quotes the following as being involved with plant interrelations involving community concepts: Heer (1835), Leoq (1854), Sendtner (1854), and Kerner (1863). Meanwhile Clements (1905) has traced the origin of the recognition of the "plant formation" back to Linnaeus (1737) but gives the credit to Grisebach (1838) for recognizing it as the fundamental feature of vegetation. Clements and Shelford (1939) state that the idea of a plant community goes back two centuries. Wherever the original credit lies, it is evident that community and association concepts were introduced by botanical ecologists and that the idea of involving animals came later. According to Clements and Shelford (1939), this joint consideration of plants and animals dates from von Post (1868). Possibly the first animal community was described as such by Möbius (1877). In translation quoted from Allee *et al.* (1949, pp. 35–36) the significant passages are "Every oyster-bed is thus, to certain degree, a community of living beings. . .. Science possesses, as yet, no word by which such a community of living things may be designated . . . . I propose the word Biocoenosis for such a community." Forbes (1887) was responsible for further development of community studies.

Quantification of such community studies apparently began with Hensen (1887) on marine plankton which lends itself to quantitative collection. This led to a flood of plankton work in which the data were quantitative but the methods of analyses were not. The development by Petersen (1914) of grabs sampling a discrete area of the sea bottom led to quantitative benthic sampling, but again quantitative data were analyzed by subjective methods. Petersen's work was the starting point of much benthic work; he initiated the concept of "bottom communities" on soft marine bottoms, and these were biocoenosis.

The concept of successions appear to have arisen from botanical studies of sand dunes by Cowles (1899) followed by zoological studies by Shelford (1907) and Adams (1909) who reviewed the development of the concept. Meanwhile marine ecologists were well aware of the temporal changes in plankton catches. Other important concepts and terms were "constancy" in an ecological sense by Brockmann-Jerosch (1907) and "coefficient of association" between species in an ecological context by Forbes (1907), who furthermore defined his coefficient in mathematical terms.

b. Developments during This Century

In designating communities in terms of their biota the botanists have again been far in advance of either zoologists or "biologists" and their work begins early this century. There are good reasons for the early rise of the phytosociologists. First, on land most plants are more obvious than most animals; second, they have more obvious effects on the abiotic environment, for example, by shade; third, because there are fewer species of plants than animals their taxonomy is better known and they are hence more easy to identify; and fourth, they are static. This means that quantitative samples are more easily obtained and the temporal factor is less embarrassing; the squirrel moves from tree to tree but other than in such fiction as Shakespeare's "Macbeth" the trees are static.

Community concepts in botany have been reviewed extensively by Whittaker (1962) and McIntosh (1967b), and three schools can be recognized before the influence of numerical methods becomes of major importance. These three schools are discussed below.

i. *Dominance–Constancy School.* In Northern Europe after about 1900 there developed a school which recognized associations by one or two species (often trees) that were constantly present and dominated the community: This has been called the dominance school—a better term might be the dominance-constancy school. There are two main objections to this approach. The first is that it is difficult to apply to complex floristic situations where numerous species appear equally codominant. (This was one reason for the development of the second school.) The second objection is that little attention appears to have been paid as to whether or not the subdominant species conform in their distributional patterns to the dominants; later evidence suggests that many do not. Expressed in different words the dominance-constancy school appears to have discarded the great amount of information which was potentially available on the less abundant species. This approaches the biotope concept in which the environment is divided and the organisms made to fit, with the difference that the obvious organisms not the abiotic factor or factors divide up the environment. It is still the approach of many ecological texts to terrestrial ecology; Odum (1971) is a recent example.

ii. *Braun-Blanquet School.* The approach and conclusions of this school have been summarized by Braun-Blanquet (1951). The school arose in southern Europe where the flora is more complex and where designation by a few dominant and constant species did not suffice. This school recognized plant associations or community types mostly by two concepts, the constancy and fidelity of characterizing species. These concepts have already been mentioned in a taxonomic context; we now consider them in a sites/species context. A species is highly constant if it appears in all the samples or quadrats within an association, but it need not be restricted to a single association. Conversely, a species is highly faithful if it occurs only in a single association, but it need not occur within all the samples within the association. Constant species are often dominants, but faithful species which have a narrow spectrum of ecological requirements are often far from obvious in the communities they characterize.

These concepts of constancy and fidelity together with dominance are important to an understanding of the numerical classifications of ecological data. They have long been used by plant ecologists, but have a much shorter history of use by zoologists and marine biologists (see Fager, 1963; Stephenson *et al.*, 1970).

The criticisms of the Braun-Blanquet school have much in common with those of the dominance-constancy school and have been outlined by Poore (1955a,b,c). They include selecting samples rather than taking them at random and sometimes making subjective judgements in the recognition of associations. The grouping of associations became too rigid and there may have been tendencies to make new data fit the accepted scheme. The greatest contribution of this school may have been the clear recognition that three criteria can be used for selecting characterizing species, those of dominance, constancy, and fidelity. The application of these concepts to animal ecology and marine benthic studies has been reviewed elsewhere (Stephenson, 1973a,b).

iii. *School of Individualistic Dissenters.* When the distributions of terrestrial plants are sampled in an unbiased way, each species tends to have its own particular distribution and one which appears not coincident with any others, or only coincident to a limited degree. Whittaker (1962) states firmly that the weight of evidence from species distributions is heavily on the side of the principle of species individuality. Later, Whittaker (1970, 1972) developed the continuum theory and showed that one might expect each species to find a different niche on an environmental gradient. If distributional patterns are completely continuous, it becomes impossible to delineate communities or associations and most difficult for the human brain to comprehend the date in totality. There seems general agreement that ordination of data, rather than delineation of groups, may be the best way of handling this type of situation (see Chapter 13).

Meanwhile, various authors have made it clear that there is not a fast dividing line between the views of the individualistic dissenters and those of the numerical analysts. Without the numerical work of the former, the existence of continua or near-continua would not have been recognized and the need for numerical analyses would have been less pressing.

4

General Comments on Classification

A. CONTINUA IN NONBIOLOGICAL SITUATIONS

In briefly reviewing the prenumerical approaches to taxonomic, biogeographical, and ecological classifications, we have tried to stress an underlying theme and this is the attempt to segregate continuous data into discrete entities. In practice this theme is less important in the taxonomy of the present-day flora and fauna than it is in paleotaxonomy and ecology.

While this book is written primarily in a biological context, it will be realized that attempts to find "satisfactory" breaks in continuous sets of data are part and parcel of many human activities. Examiners of every kind of human attainment know that the marks obtained usually approximate to a normal or some other unimodal distribution curve. Also every examiner knows that across the continuous spectrum we have our categories of credit, pass, and fail—or whatever we may substitute for them. We have to learn to live with the concept that say 49% is a failure and 51% is a pass. The problems of segregation within a continuum or near-continuum are of particular concern to all the social sciences in which quantification is involved whether they be educators, psychologists, anthropologists, or sociologists. It is no accident that many of these disciplines have contributed to numerical analysis.

The manner in which nonbiologists have approached the classification of three separate continua will serve to stress the universality of the problem and of the need to develop techniques for recognizing groupings within continua. Crystallographers are frequently confronted with the need to classify isomorphic mixtures of minerals and usually subdivide the continua on the amount of each member present in the mixture. The classification of plagioclase felspars as given in Table 4.1 is an example of this approach.

Long before physicists subdivided the electromagnetic spectrum into x-rays, light, heat, etc., the layman had recognized and given names to

TABLE 4.1

THE CHEMICAL COMPOSITION OF MEMBERS OF THE ALBITE (Ab, $Na_2O \cdot Al_2O_3 \cdot 6SiO_2$) ANORTHITE (An, $CaO \cdot Al_2O_3 \cdot 2SiO_2$) SERIES OF PLAGIOCLASE FELSPARS

Member	% Anorthite
Albite	<10
Oligoclase	10–30
Andesine	30–50
Labradorite	50–70
Bytownite	70–90
Anorthite	>90

certain regions of the visible spectrum and so the names of colors were born. As the third example we refer to the recognition among sound waves of a series of notes defined largely by pitch and written within the framework of a series of lines with intervals between them.

B. WHAT CLASSIFICATION INVOLVES

The term "classification" requires amplification (see Cormack, 1971) becauses it encompasses at least two other concepts, those of identification (Dagnelie, 1966) and dissection (Kendall and Stuart, 1966). Identification is the process of choosing as to which of a number of defined classes a new entity should be allocated. However, in many cases, both taxonomic and ecological, admission of a new entity may alter the class and necessitate redefining its characteristics, so that it becomes an integral part of the classificatory process rather than something quite separate.

Dissection is the splitting of a continuous into a discontinuous series, a process that has been stressed earlier. Because in many of the situations we are concerned with there is likely to be merging between completely continuous, near continuous, and completely discontinuous data, it seems that dissection should not be treated as something separate from classification. Dissection is allied to group formation in that it leads to the definition of groups in otherwise continuous distributions. Here the decision as to where to fix the boundaries between groups is left to the user of the system. The previous example of grading examination results into classes of pass and fail is an example of dissection. Sometimes the original data suggest the sites for dissection, at other times the user must rely entirely on his discretion and the purpose for which the classification is being made.

Identification and dissection are only two aspects of classification; there are many kinds and many aspects of classification. In what follows we will touch on most of these but not deal with them in depth. The interested reader is referred to the expanding literature devoted to classifications of classifications and to reviews and comments on them.

It may be noted there is an increasingly mathematical content in this literature and it is important that the models and principles on which the mathematics are based should be biologically well founded and the conclusions reached tested against further experience. In the majority of instances there are no absolute criteria against which to test the structure of a classification once derived and so it is important to be clear about the steps taken in its derivation.

Recent papers on classificatory strategies include those of Macnaughton-Smith (1965), Kendall (1966), Good (1965), Lance and Williams (1967a), Jardine and Sibson (1968, 1971a,b), Sibson (1971), Williams *et al.* (1971a, b), Cormack (1971), and Williams (1971., 1972).

Lance and Williams (1967a) in listing four types of classifications included simplification of the data by ordination. Allied to classification, the process of ordination may or may not lead to the recognition of groups within a series of sites or taxa. Prior to ordination the entities being considered (e.g., taxa or sites) are assigned to positions in a multidimensional space defined by their properties or some measure of their dissimilarity. Efforts are then made to express the relationships between the entities in fewer dimensions than those originally considered. We treat ordination as being virtually separate from classification and accept that a much fuller treatment might have been desirable. To do so, however, would have expanded the text to an unwieldy size.

Williams (1971) uses a dichotomous key to classifications, and his first choice is between overlapping or nonexclusive and nonoverlapping or exclusive. In the former an entity may appear in more than one group. Such classifications are widely used in libraries. Examples, quoted from Lance and Williams (1967a), include Needham (1962) and Needham and Jones (1964). In general taxonomists have not favored such classifications though Bor (1960) placed several of the same species in both the genus *Saccharum and Erianthus*. Nonexclusive classifications have also been employed occasionally by ecologists (Yarranton *et al.*, 1972).

Williams (1971) divides nonoverlapping classifications into *extrinsic* and *intrinsic*. In the former, the classification of intrinsic data is arranged to reveal as closely as possible the extrinsic situation. It has been particularly applied by Macnaughton-Smith (1965) in sociological analyses in which "irrelevant" resemblances may mask investigations of basic causality. Extrinsic classificatory programs are not well developed and are referred to later (see Chapter 11).

Intrinsic classifications are used to derive groups solely from their attributes. After classification the groups may be examined to determine if they reflect extrinsic discontinuities, and these examinations are particularly applicable to ecological classifications. For example, in marine benthic situations no data are given in the classificatory procedure on any site attributes other than the species. Information on substratum, depth, salinity, and season are not included among the data to be classified and enter the discussion to "explain" the site groups which are derived. In practice this hidden data can be of great value in deciding the overall "sensibleness" of different methods of classification (see Chapter 9, Section A).

Intrinsic classifications are of two types, viz., *nonhierarchical* and *hierarchical.* Williams (1971) states:

> "A hierarchical strategy always optimizes a *route* between the entire population and the set of individuals of which it is composed. . . . a hierarchical fusion strategy must necessarily optimize some function between the two groups next to be fused . . . or between these two groups jointly and their fusion product . . . . As a result the groups through which the process passes are not themselves necessarily optimal; the best route may be obtained at the expense of some slight reduction in homogeneity of individual groups. With a non-hierarchical strategy, on the other hand, it is the structure of the individual *groups* which is optimized, since these are made as homogeneous as possible. No route is defined between groups and their constituent individuals, so that the infrastructure of a group cannot be examined. Similarly, no route is defined between groups and the complete population; such a route can usually be devised, but it will simply be a 'key,' and can not be used to indicate relationship. Despite these advantages, for those applications in which homogeneity of groups is of prime importance the non-hierarchical strategies are in principle very attractive; unhappily their current state of development lags far behind that of their hierarchical counterparts, which at their best are more flexible, provide a wider range of facilities, are numerically better understood, and are computationally faster."

For the taxonomist in particular, hierarchical classifications are attractive in that they are both traditional and have a parallel in evolutionary theory. Furthermore hierarchical classifications bring organization to a body of knowledge and if science is defined in terms of organized knowledge they are the more "scientific" of the alternatives. Hierarchical nonoverlapping classification produces groups, hereinafter termed *clusters,* whose relationships to one another are readily expressed in two dimensions, generally in the form of a dendrogram.

In both taxonomic and ecological work it is often difficult to know in advance how many groupings may be required from the original data. It seems this can best be decided by a process of trial; typically it appears best to generate an excess of groups and fuse some of these later. Freedom to move from one to another number of groups is one of the advantages of a hierarchical strategy.

With hierarchical classificatory strategies the groups arise as a consequence of the methodology adopted to establish the hierarchy and do not necessarily exhibit the same homogeneity. In contrast nonhierarchical methods can produce clusters of defined heterogeneity but do not link them together in any systematic framework. The techniques are, as stated, relatively undeveloped but their use in marine ecology merits especial consideration. Fager (1957) and later Fager and McGowan (1963) have initiated a nonhierarchical method of species/species classification whereby recurrent species groups with defined characteristics have been obtained. These have been widely used in the United States of America (see Chapter 6, Section B, 1, b for listing) and part of their attraction is clearly in their prescription which ensures that groups derived by different workers are comparable.

Williams (1971.) finally divides hierarchical methods into *divisive* or *agglomerative* and either *monothetic* or *polythetic*. These concepts have been briefly introduced earlier, and will be developed in later sections. The relationships between these methods are shown diagrammatically in Fig. 4.1.

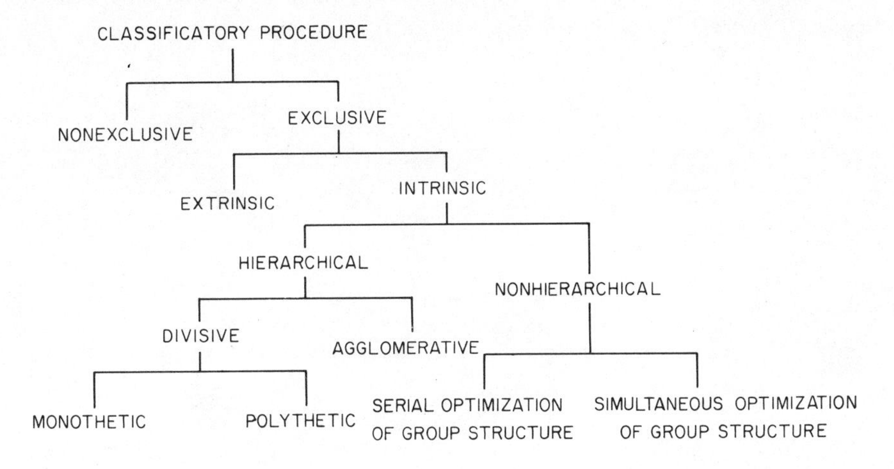

Fig. 4.1 The relationships between classificatory procedures expressed in the form of a dichotomous key (based on Williams, 1971).

In the above discussion the process of classification has been considered to involve the organization of individuals into groups. To many, however, the term classification means the placing of an individual into its correct position within an existing classificatory system—this is the process of *identification.* An alternative term is *discrimination.*

It is clear that the whole subject of classification is complex. It depends on both the nature of the variation within the system to be classified and the use to which the classification is to be applied. Accordingly it is necessary to define carefully the terms we are using before proceeding and this we now do. In what follows we endeavor to be consistent in our use and application of the following terms, as employed for numerical studies:

1. *Clustering*—the formation of nonoverlapping groups defined by hierarchical or nonhierarchical methods.
2. *Clumping*—the formation of groups defined by their internal properties and not necessarily defined with respect to other clumps.
3. *Dissection*—the formation of nonoverlapping groups defined by fixing boundaries in a continuous distribution.
4. *Identification*—the location of an individual within a group previously defined by one of the above classificatory procedures.
5. *Ordination*—the disposition of individuals in a reduced space the original of which was defined by a series of axes corresponding to the number of properties studied for these individuals.

In passing it ought be noted that the relationships between groups obtained by clustering, clumping, or dissection may be further investigated by considering an ordination of their centroids.

5

Numerical Approaches to Classification

A. INTRODUCTION

To many biologists the publication of Sokal and Sneath's (1963) "Principles of Numerical Taxonomy" was the first introduction to this subject and much of the philosophy that lay behind it. It is now evident that it was a taxonomic exposition of widespread developments which had involved, for example, psychology and botanical ecology. If followed directly from work by Sokal and Michener (1958) in zoological taxonomy and that of Sneath in bacterial classification. It appears that the book was written with a modicum of revolutionary zeal, and that some of its hopes have not been, and may never be, fulfilled.

Several interesting predictions are quoted in Sokal and Sneath's book, one made by Ehrlich (1961) to the effect that data-processing equipment will be the most important tool of the taxonomist in 1970 is unfulfilled. Another by Fein (1961) that by 1975 at least one botany department in a large university will offer a course in the application of computer techniques to taxonomic botany has been substantially fulfilled. In 1970 one of us (H.T.C.) gave a postgraduate course in numerical analysis in taxonomy and ecology in the University of Queensland. By 1973 we have collectively or individually given courses in six universities in Australia and the United States.

Apart from predictions there is an interesting near contradiction. Early in their book Sokal and Sneath (1963) state: ". . . the outstanding aims of numerical taxonomy are *repeatability* and *objectivity* . . .. Classification must be freed from the inevitable individual biases of the conventional practitioner of taxonomy." Later, they state: "It is, of course, inevitable that some of the statistics will have to be taken on trust. Few biologists have the interest in mathematics that would lead them to familiarity with all the arguments involved here. Nevertheless, they have a check on the statistics in their own shrewd judgements on the worth of the results." Numerical classification is partly open to the very criticism it seeks to avoid, the subjective element, and at a later stage the partial safeguards against this are outlined. While Jahn's (1961) hope that "intellectronics" might serve as an extension of man's intellect seems misplaced, it is evident that numerical analyses can extend our powers of data analysis, particularly in the ecological field. In the taxonomic field the overcaution of Michener and Sokal (1957) seems more justified than the overenthusiasm of Sokal and Sneath (1963). The former stated: "We have no belief that our method will replace standard systematic procedure, although we believe it may frequently be useful to check such procedure by a statistical approach."

Following the pioneer work of Sokal and Sneath (1963) the next landmark is the book by Sneath and Sokal (1973). Of the same high standard as the first volume, the second supplements the diverse origins of numerical taxonomy and makes reference to most of the more recent numerical taxonomic literature. The original revolutionary zeal still shows, though there has been some softening of attitudes, for example, with respect to the weighting of attributes. In this volume we hope that our own zeal shows through and that the role of numerical classification in ecology as well as in taxonomy is made clear.

In the next sections we consider the meat of the situation, including types of data, methods for estimating similarity between individuals and/or groups, and methods for making classifications. Relative merits of alternative procedures will be discussed but we should warn that rarely will it be possible to say which of them is "best." This will depend on both the nature of the data and the purpose for which the analysis is being undertaken. In a relatively simple study, it may be appropriate to use several measures, for example, when considering the problem of size in human beings one might measure any of the following: weight, height, volume, or girth but no one of these is "best" except perhaps in a particular context. In the more complex studies we shall be considering the judgment of "best" as this needs particularly careful consideration.

B. TYPES OF DATA

1. General

We preface discussion of different forms of data by consideration of the amounts of data which are required for effective numerical classification. It is perfectly possible to classify individuals by a single attribute. Consider the following example of recordings of a single species in six sites.

	Sites					
	1	2	3	4	5	6
Species A	1	2	1	50	60	55

The existence of two site groups here is obvious by the simplest of numerical inspections and this would be too trivial an exercise for "proper" numerical classification. If in the above example the number of different species was increased to about six and the data throughout were less obviously discontinuous, one would reach the approximate limit of immediate visual analysis; in other words, a 6×6 data matrix using whole numbers. With only presence/absence data possibly a 10×10 matrix can be scanned visually.

Beyond this, the more elaborate methods that we outline are likely to be necessary. The next question is how many attributes are desirable to obtain a "reliable" result. This is important because to a greater or lesser extent the similarity between entities depends not only on the attributes measured but also on their number. There is no precise answer to the above question, for the number appropriate is a property of the data themselves. Whenever the entities to be classified associate into relatively homogeneous groups, with marked discontinuities between the groups, a relatively few attributes will suffice to generate a reasonable classification. In contrast if the data are less well structured, the groups being less homogeneous and not markedly different from one another, many attributes will be required to achieve a reasonable classification.

Thus in the absence of a theoretical basis for determining the number of attributes to be measured, the problem must be approached empirically. For example, in taxonomy, having classified a set of entities on the basis of a set of attributes, the number of attributes may be increased and the set of entities reclassified. Should the two results be similar the classification may be regarded as stable, if not, further attributes may be considered until a stable classification is achieved or the objective is abandoned.

Data which can be obtained in taxonomic and ecological work show many features in common, but there are some differences in emphasis which affect the ways in which these data can be used. The similarities and differences are outlined below.

One overall difference is in the nature of the characters or attributes which are used to classify whatever it is to be classified. In taxonomy in the usual classification, species are classified and the attributes employed are generally their structural features. In ecology in the usual classification of biota data, sites are the entities and the attributes are the species found thereat.

It is possible in both taxonomic and ecological classifications to classify the attributes by the entities. In taxonomy, this shows resemblances between structural features and the approach does not appear to have been followed in depth. In ecology, the sites of occurrence become the attributes of the species, and one determines species groupings. These have considerable value and allow us to have associations of species to compare or contrast with associations of sites. As stated earlier (Chapter 3, Section B,2), classifications of attributes by entities are now usually referred to as "inverse classifications" in contrast with "normal" ones.

Previous terminology distinguished between these classifications by referring to entity classifications as a "Q" and attribute classifications as an "R," respectively. The use of these letters should be avoided for they do not readily indicate meanings and sometimes these have been reversed (as discussed, for example, by Ivimey-Cook *et al.*, 1969). In the account which follows, classifications may be considered as normal unless otherwise specified. That is, taxonomic classifications associate species in terms of their features and ecological classifications sites in terms of their species.

We should reiterate that, in ecology, classifications may be based on abiotic attributes of the sites as well as on the presence or absence of species or their abundance. For example, terrestrial sites may be classified in terms of such attributes as annual rainfall, soil depth, bedrock, slope, or aspect and marine sites by depth, salinity, oxygen concentration, pH, light penetration, etc. The same techniques, albeit with some prescribed restraints, can be used with nonbiotic data on a wide range of broadly ecological situations, for instance the classification of streams or wells can be undertaken solely in terms of their abiotic attributes or the classification of regions in terms of their climatic data.

2. Kinds of Attributes

Data consist of attribute scores, and clearly it is necessary to consider the different kinds of attributes and their interrelationships.

In general, attribute data are of six basic types:

(i) *Binary*—possessed of two contrasting states, such as the presence or absence of a species in a quadrat, or fruits being dehiscent or indehiscent.

(ii) *Disordered multistate*—possessing three or more contrasting forms each ranking as equal. For example, flower color red, blue, or white.

(iii) *Ordered multistate*—possessing a hierarchy of contrasting forms which encompass the total variation in the range of entities under study. For example, organism abundance in quadrats might be described in terms of the series rare, common, or abundant; plants might be grouped according to whether they developed short, long, or very long fruits. Attributes such as these are also known as *ordinal or graded.*

(iv) *Ranked*—as for ordinal except in that the grading applies only within a single entity. Ecological examples are the ranking of species within a given sample by whatever ecological criterion has been decided on. These criteria could include such measures as cover or dominance. Because the scoring of the attributes applies to single entities only, data are site-standardized (see Chapter 7, Section D). Ranking of attributes in this sense is rarely practiced by taxonomists.

(v) *Meristic*—possessed of several values each of which is a whole number. For example, the number of anterolateral teeth on the carapace of a portunid crab or the number of petals possessed by a flower.

(vi) *Continuous*—measures of size on a continuous scale as with attributes such as length or weight. In practice this approaches meristic in that the accuracy of measurements of the scale restricts the number of decimal points which are employed.

Besides the above attributes, there are some of which we have little experience. These include a category of *nonexclusive multistates* such as arise in the study of multicolored parrots where color combinations are involved as well as the basic colors. Thus some birds may have red and green plumage, others red and blue, and yet others red, blue, and yellow.

Generally speaking taxonomic studies tend to be characterized by attributes capable of expression as binary, disordered multistate, or meristic data while ecological studies can involve the use of a wider range of data forms.

The attribute categories are in fact, less separate than they might appear and depend in large measure on the sampling procedure adopted. For example, an attribute scored as binary in terms of single sample sites, may be better expressed in meristic form if the sites are subsequently considered as clusters. Consider, for example, a series of sites sampled at regular intervals along a line transect. It may be decided after sampling to group the sites into clusters of three before proceeding further with the analysis. A given species may now be present in none, one, two, or three of the sites and so its amount in the cluster could be scored as 0, 1, 2, or 3. Likewise, when data resulting from the scoring of meristic or continuous attributes are standardized by entity (see Chapter 7, Section D), they resemble data arising from rank-scored attributes. We now consider the six types of data in more detail.

3. Binary Data

a. In Taxonomy

Binary attributes are of two kinds, the presence or absence of some feature being one and the form of the same feature being the other. In essence these are quite different in character, a difference which it is important to appreciate.

Consider a presence–absence attribute. Here "presence" may lead to secondary or further attributes and such situations are very common in taxonomic studies. Thus in comparing beetles the presence or absence of elytra might be recorded, and for those insects with elytra further "consequential attributes" arise referring to the properties of the elytron. There are thus more characters available for comparing insects with elytra than for comparing those without them. Furthermore these are available on a single comparison between insects possessing and those lacking elytra. The statistical consequences of such situations have been explored by Kendrick (1965), Williams (1969), and McNeill (1972) and are best understood against a background of the similarity measures available.

A number of methods have been proposed to handle this situation but none is entirely satisfactory. One is to weight the original binary attribute to equate with the number of consequential attributes generated by the possession of the primary attribute. This has dangers which can be illustrated as follows.

Beetles may either possess or lack elytra. For those with elytra, the further attributes elytra smooth or punctate and elytra long or short are available. Using the weighted method a beetle with smooth short elytra would have a greater apparent resemblance to one lacking elytra than to a beetle with punctate long elytra (see Table 5.1A). There would appear to be no logical justification for this viewpoint.

Alternatively, nonconsequential attributes (i.e., absent because the primary one was "negative") may be treated as missing attributes (see Table 5.1B). This is somewhat different from real missing attributes, which we shall now consider. These can arise if only a single specimen is available, particularly if it is a type which would not be dissected. (In theory, following the "New Systematics," one should always examine populations, but the practicing taxonomist knows this is often impossible.) It can also arise in botanical taxonomy if important structures, for example, flowers or seeds, are not available in the samples. There are two solutions, regarding missing attributes, either to eliminate completely those which are occasionally missing or to accept that in some interspecific comparisons a smaller

TABLE 5.1

TWO METHODS OF TREATING CONSEQUENTIAL ATTRIBUTES

(A) Basic attribute weighted according to the number of consequential attributes[a]

	Attribute	Taxon A	Taxon B	Taxon C
Basic	Elytra present	+ +	− −	+ +
Consequential	Elytra long	+	−	−
	Elytra punctate	+	−	−

(B) Consequential attributes treated as missing attributes (*)[b]

	Attribute	Taxon A	Taxon B	Taxon C
Basic	Elytra present	+	−	+
Consequential	Elytra long	+	*	−
	Elytra punctate	+	*	−

[a] Even with basic attributes doubly weighted, taxon C resembles taxon B as much as it does taxon A.

[b] Taxon B comparisons involve only a single attribute, and this taxon differs completely from taxa A and C.

number of attributes are used in comparisons. The choice might depend on the method of making comparisons, as discussed later.

While in the previous example it appears sensible to treat the absent consequential attributes as missing attributes, yet in some cases this does not resolve all the difficulties, as illustrated by Table 5.2. Here we have three taxa (A, B, and C) and seven attributes (1–7), with the first attribute generating three further ones (2, 3, and 4). In terms of overall similarity the data indicate a greater resemblance between taxa A and C (3 attributes out of 4) than between taxa A and B (5 out of 7).

A third solution to consequential attributes is to treat them as being multistate attributes, and in the earlier example we could thus have such alternatives as "elytra absent" "elytra present and long," and "elytra present and short." This also involves problems; it assumes that there is no mutual dependence between states, which is clearly untrue.

One should stress that, in taxonomic data, there is a marked distinction between a "true negative" response to a question regarding attributes, a negative due to a missing attribute, and one due to a nonconsequential attribute. Similarities between two taxa can often be shown just as much

TABLE 5.2

THREE TAXA SCORED FOR SEVEN ATTRIBUTES, THREE OF WHICH, ATTRIBUTES 2–4, ARE CONSEQUENTIAL TO ATTRIBUTE 1

Attribute	Taxa		
	A	B	C
Basic			
1	+	+	−
Consequential			
2	+	+	*
3	+	+	*
4	+	−	*
Basic			
5	+	+	+
6	+	−	+
7	+	+	+

[a] +, attribute present; −, attribute absent; *, no score possible because of nonconsequential attribute.

by a series of paired true negatives as by a series of positives—in other words "double negatives" must count.

Irrespective of whether presence or absence attributes involve consequential attributes it cannot be assumed one state is more important than the other. For instance in the portunid crabs, some species of the genus *Portunus* have three blood-red spots on the carapace, others none. Because this attribute is conspicuous it appears early in some keys to the genus, yet it cannot be held that the attribute is particularly important because more *Portunus* species lack such blood-red spots than possess them.

With binary attributes in which both states are presences (such as flowers yellow or blue) there is, as a rule, no reason to regard either state as more important than the other.

b. IN ECOLOGY

Here we include biogeography, in which the bulk of the data have been binary and are concerned with large-scale patterns, and also ecology in a more restricted sense, which is concerned with smaller scale patterns. In practice many studies are roughly intermediate (see, for example, Wolff, 1970).

Differences of opinion exist on the value of binary data in ecological work, but the consensus of opinion seems to be that other data are preferable. For example, Greig-Smith (1964, p. 160) states in a botanical context: "We are, in fact, dealing with a population of individuals (if stands may be so regarded) which differ from one another in terms of continuous variables, of which presence and absence are only a crude expression. With only slight exaggeration we may, within the limits of a set of broadly similar stands, regard absence as simply the extreme value of a continuous variable." Earlier in a marine benthic context, Petersen (1914) criticized the use of presence/absence data and fauna lists, stating: ". . . the long lists just obscure the differences especially if not accompanied hy (*sic*) quantitative information. . .."

In most branches of ecology the tendency is to regard dominance as important, whether measured in grades or ranks or numbers or biomass or energy flow. As we shall see later the results of analyses using data with numerical values are more informative than those using binary data. To continue to use binary data in ecological surveys is justifiable only on two premises. The first is difficulty in obtaining anything else, for example, when inordinate time might be required to obtain numerical values. The second is that, in advance, there is a declaration of lack of interest in dominance, with attendant emphasis on such concepts as ubiquity, constancy, and fidelity of species.

A somewhat contrary view has been expressed by Lance and Williams (1967b) who state: "It is already known (Lambert and Dale, 1964) that in the case of the matrix of continuous quantities with many zeros, dichotomising at the zero/nonzero boundary loses little information; and in the more general case we have ourselves shown (El-Gazzar *et al.*, in press) that, although information is undoubtedly lost by dichotomisation, the extent of loss is less than might intuitively be expected." The reference is to El-Gazzar *et al.* (1968). We stress that this applies to matrices with many zero entries and that in such cases it is common to eliminate rare species and reduce recordings of zeros. We would also stress that the "Canberra school" has altered its views. In a recent trial of a variety of dissimilarity measures, Williams *et al.* (1973) noted that while in a simple situation they studied involving eight sites only presence/absence was required, for 10 sites there was "some advantage" in using numbers, while for 80 sites (closest to an actual primary survey) quantitative data were distinctly preferable.

Assuming that presence/absence data have been obtained, some of their shortcomings should be appreciated. Some of these concern the species which occur infrequently throughout and the stress to be given them (see below). Others concern the boundaries of specific distributions; binary data give undue prominence to the extremes of the range of a species, where elements of chance may largely determine whether or not a species appears in the samples.

In most ecological survey work, a few species will be obtained on many occasions and many species on only a few occasions, including once only. The unique occurrence of a rare species at a single site is unlikely to characterize that site in any meaningful way, and in benthic work it has been found desirable to eliminate such species (Stephenson *et al.*, 1970, 1972; Stephenson and Williams, 1971).

The importance of species occurring more than once, but still infrequently, requires careful thought. In work on plankton, with a history of the importance of indicator species (Russell, 1935, 1939; Fraser, 1952, 1955; Mankowski, 1962), it might be desirable to retain them in the analysis. Most benthic workers have eliminated from their analyses the species which occur infrequently, as have some recent nektonic workers. (See Chapter 7, Section B.)

Another way of approaching rarer species is the decision on whether or not paired absences should count in the later analyses. If rare species are included, the analysis can be dominated by their conjoint absences; if on the other hand joint absences are completely excluded, this can obscure important conjoint absences of ubiquitous species. No invariable rule can be prescribed, and the only solution is to try a variety of methods and determine which "comes out best."

When using biogeographical and ecological data, direct problems over consequential attributes do not appear, although the concept might merit consideration. The presence of certain organisms quite frequently results in environmental conditions in which further species can also be present. Apart from the more obvious host and parasite or host and commensal relationships this occurs in tropical waters with coral reef and mangrove biotas. In actual analyses these situations would show clearly and not cause mental confusion as they may do in a taxonomic context.

Consequential attributes could also arise in ecology if the different life history stages are recorded as well as the presence or absence of the organism. Sometimes there are good reasons for this—differences between the locations of seedlings and adult trees may be quite meaningful. In other circumstances the absence of one life history stage could be entirely consequential on the absence of another, and this is exemplified by the practical control measures of many of the parasites of man.

Missing attributes do not appear to be a serious ecological problem, but do not arise in practice. A familiar example is the failure to preserve adequately one of a series of samples. Another arises through the confusion of species early in a set of samplings and their subsequent elucidation. Unless all specimens in the early samplings have been retained, there is no way in which the data can be recovered, and all records in the confused groups are likely to be discarded in a total analysis. It will be evident that these are not problems of binary data alone.

As stated previously, presence/absence data in ecology become meristic when one compares the number of presences in a variety of summated situations. Greig-Smith (1964) has discussed certain of the older and better known forms of comparison. For example, he quotes Raunkiaer (1934) and notes that from presence/absence data from numerous small samples within a community one can readily derive percentage frequency. Greig-Smith also notes there are relationships between frequency and density, but that these are difficult to establish either empirically or theoretically because frequency depends on the pattern of distribution as well as the density. This raises the complex question of quadrat or sample size, which is best avoided at the present stage.

Work by Stephenson *et al.* (1970) exemplifies a method whereby binary data can be converted to meristic. In a study of dredged benthos of Moreton Bay and following a method devised for rain forest work by Williams *et al.* (1969), the presence/absence records of a site were combined with those of the five nearest-neighbors and the data on the site clumps were classified. It was stated (Stephenson *et al.*, 1970, p. 476): "There are three advantages of this procedure: first, the rare species are over-sampled in a systematic manner. This is comparable to standardization by variance in a Euclidean model . . . . Secondly, the method recognizes as associations two species which occur in neighbouring sites, which is realistic when sampling the present environment. Thirdly, it smoothes out minor local deviations from the overall pattern." Use of "clumping" is not restricted to binary data. In a later study of the grab benthos of Sek Harbour in New Guinea, Stephenson and Williams (1971) again used site clumps in the analysis, this time using numbers of individuals instead of their presence and absence.

4. Disordered Multistate Data

These are used widely in taxonomy and can be analyzed in either of two ways. The first is to convert them to a series of binaries, for example, flowers "red," "white," or "blue" can be resolved into three sequential presence–absence alternatives which are "red" and "not red," the latter leading to "white" and "not white" (blue is "not white").

Under these circumstances when flowers are considered in pairs they exhibit one of two levels of similarity. They are alike with respect to all three binaries or with respect to one only of the three. For example, a pair of red flowered plants will be alike not only in their possession of red flowers but also in their lack of blue or white flowers. In contrast a white- and a red-flowered plant will be alike only in that they are not blue. As the number of states available increased for a disordered multistate attribute so any pair differing from another appear to become more alike for they always agree on sharing the attribute states they do not possess.

The second method of analysis is to offer each entity a choice of one of n scores where n is the number of states assumed by the attribute. Here the degree of similarity is measured in terms of dissimilarity by using information theory, for example, to estimate the increase in diversity resulting from combining the two elements. (See Appendix.) Clearly the less dissimilar the entities are the more similar they must be.

5. Ordinal, Ordered Multistate, or Graded Data

Of the forms of quantitative data, these are the easiest to think about, and many forms of numerical data are reduced to this form during or after analysis before presentation to and assimilation by another person. If a person with some experience of intertidal ecology is told that acorn barnacles are abundant in one situation and rare in another it immediately means something, whereas to be told there are $5871/m^2$ compared with $43/m^2$ may mean a great deal or very little.

The ease of appreciation of this data applies not only at the level of data presentation, but also during data collection. With reasonable experience, categories of abundance in ecology or relative sizes in taxonomy can be appreciated at a glance. To many biologists it has long appeared a waste of time to collect more complex data merely to convert it back to the same mental impression in the end. This concept of wasted effort, added to the difficulty in understanding numerical methods, has been one of the reasons why these methods have only partially and slowly become accepted.

There are difficulties in obtaining the correct ordered multistate values and these are listed below in an ecological context.

(i) By intuition, the observer converts a continuum, or near continuum of data into an ordered form. That is, he creates discontinuities where they doubtfully exist. This leads to the probability that different observers have different grading standards depending on their backgrounds of interest and experience. A field worker who is investigating a general situation is often astonished when a group expert using his collections describes a species as "common." The field worker uses "common" against the background of the remaining biota, the group specialist's background is possibly the number of specimens of the species accrued in the museums of the world.

(ii) There may be considerable personal factors in assessing grades. Greig-Smith (1964) has said: "Hope-Simpson (1940) has shown that one observer may give markedly different assessments on different occasions, particularly at different seasons. Smith (1944) . . . showed that individual observers out of a group of eight deviated in their assessments from the group mean by as much as 25 per cent. . . . A further source of personal error lies in the mental state of the observer. Every ecologist with some experience of frequency estimation is aware that rare and inconspicuous species tend to be rated lower when the observer is tired than when he is fresh and alert." One should note that some of these criticisms are not restricted to grading assessments, even in counting specimens, inconspicuous individuals tend to be overlooked in adverse conditions.

(iii) A third problem is to know whether the grades are based on raw or transformed data. While we know of no comparisons of grading systems in different disciplines, there has been acceptance of grading on a logarithmic basis by a number of marine biologists. These have included intertidal studies by Crisp and Southward (1958) and by Southward (1962, 1967), sublittoral benthic studies by Field (1970), and plankton studies (Anonymous 1973, and references therein). In many of these cases the grading was used to simplify data which were available in meristic or continuous form, prior to analysis. This is different from postanalytical grading which is considered in Chapter 10, Section D.

Meanwhile in ecology it seems that when dealing with thousands one thinks of "abundant," with hundreds of "common," and with tens of "present." There is no great harm in dealing with a system which has an intrinsic logarithmic conversion provided its existence is realized.

(iv) The fourth difficulty is in the differing numbers of grades which can be used. In comparing similar biotas with a strictly comparable technique and using the same trained observers throughout as many as six grades can often be employed, for example "abundant," "very common," "common," "present," "rare," and "not observed." In different circumstances as few as three grades may be used (Stephenson *et al.*, 1970).

(v) A fifth difficulty is the ecological basis of grading, for example, by numbers of specimens or by numbers multiplied by a size assessment. The problems which arise are well illustrated in the reanalyses of Petersen's (1914) classic data by Stephenson *et al.* (1972). Petersen counted animals and weighed them, but in his description of communities he used characterizing dominant animals, with the dominance based on numbers *and* weights. Stephenson *et al.* (1972) have shown, as might be expected, that there is little correspondence between the classification resulting from the use of numbers and that resulting from the use of weights. It seems dangerous to attempt to grade on a numbers *and* weight basis, and that a choice should be made. Nonetheless it should be noted that a numerical analysis in ecology by Loya (1972) on Red Sea corals uses as biotic data two entries per species—number of individuals and living cover. Justification for this action is neither given nor is it apparent.

Greig-Smith (1964) has discussed similar problems in terrestrial plant ecology. Considering grading by the use of frequency symbols he states

> "Several factors influence the observer in his assignment of a frequency symbol. Those uppermost in the minds of most observers are probably *density* or number of plant units per unit area and *cover* or percentage of the total area covered by aerial parts of plants of a species, rather than true frequency, which is itself a complex character (see below). It is, however, difficult to avoid being influenced by the differing growth forms of different species and by the varying pattern of distribution of the individuals of different species on the ground, two factors greatly affecting the relative conspicuousness of different species. Even if density and cover alone are taken into consideration, an ideal probably impossible to attain, an attempt is being made to assess on one scale two largely independent variables."

Because of difficulties outlined above, potential use of ordered multistate data in ecology should be the result of careful deliberation. It will be appreciated that some of the same problems apply with equal force to taxonomic data.

6. Ranked Data

As already indicated the only difference between ordered multistate and ranked data is that the rankings have comparative value within the sample, but not necessarily between samples. In ecological surveys it is easier and sometimes more meaningful to work within samples. For example, when the performance of sampling gear depends on the sampling conditions, it is difficult to obtain ordered multistate data, and the number of grades which can be recognized is severely restricted. Provided one is prepared to work on a "within sample" basis many more ranks can be recognized or by conversion to percentages the data can become continuous. This procedure has been used by Hailstone (1972) in analyzing dredge catches, and has the same advantages and disadvantages of using site-standardized continuous data. We shall return to this in a later section.

The difference between ranking and grading is often indicated by the vocabulary; in the former such terms as "dominant" and "subdominant" are used while in the latter the terms include such things as "abundant" and "common." It is well known that a dominant form may often be far from abundant—desert shrubs and gastropods high on intertidal rocky shores are examples.

Greig-Smith (1964) has rightly criticized vocabularies including "dominant" and points out that there are three distinct ecological meanings of the term. Usually it means (in a botanical context) the highest grade of density plus cover of the vegetation. To those ecologists who accept an organismal view of the community, dominance may imply degree of influence exerted over other species of the community—in this context man may well be dominant over many "dominant plants." A third botanical use applies to the tallest plants in a community.

7. Meristic Data

These are in the form of whole numbers. Clearly they closely resemble continuous data, and equally clearly continuous data restricted to a certain number of significant figures can be converted to meristic form by multiplying by a power of ten, or rounding to the nearest unit.

Taxonomic data are often meristic in form, for example, there are 9, 8, 7, 6, 5, or 4 anterolateral teeth in portunid crabs. There is an appealing but deceptive precision with this type of data. Thus in the above case, there are intergrades between the 5 anterolateral teeth of the genus *Thalamita* and the 6 teeth typically found in the genus *Charybdis*. The difference hinges on whether a small projection which occurs in some species between two of the teeth is or is not a tooth.

In ecology when dealing with numbers, the bulk of the data are meristic, but problems arise on and near the boundaries of quadrats.* Here only

* Although recognizing that there are differences in their meanings, we will generally treat "quadrat," "sample," "site," and "station" as being synonymous and implying "sampled area" or "sampled volume." In fact there are ambiguities about "sample" and possibly "station." This is because they may imply an area that is sampled at a particular time. Later, the importance of the distinction between the two will be discussed. Meanwhile "traverse" is generally taken as "sampled line" or "transect" although in practice it does have breadth, and may consist of a linear arrangement of quadrats.

part of an organism may be present in a quadrat, and clearly the larger the quadrat in relation to the size of the organisms which are being sampled, the smaller the problems of dealing with meristic data. Sometimes the difficulties can be resolved by neglecting all cases in which less than half an organism is present—this has the effect of impoverishing the boundaries of the quadrat.

8. Continuous Quantitative Data

In taxonomic studies measurements of lengths or ratios of lengths produce a common type of continuous data. In other cases data that are meristic with respect to individuals become compatible with continuous when populations are considered and the meristic data are expressed in terms of means, standard derivations, ranges, variances, etc. If the meristic data are transformed, for example, by square roots, they again become continuous. Similar considerations apply to meristic ecological data.

Later we shall contrast polythetic and monothetic classifications in detail (see Chapter 8, Sections B and C). At this stage we should note that any system involving the polythetic comparison of numerous entities with numerous attributes each involving means and variances could be extremely complex. One of the ways in which a standard statistical approach can be married to an approach by numerical analysis is to use data such as means and variances of attributes to dissect a single attribute into several. It is well known in ecology that the distributions of different sizes or age groups of a species may differ, and to preserve this dissimilarity there can be no objections to regarding the different size groups as different attributes. Years ago, Petersen (1914) visually distinguished attributes in this way with separate listings for example of adult and juvenile *Venus*. Williams *et al.* (1973) incorporated different size classes of rain forest trees into their numerical analyses; their conclusion (p. 68) is that these "should *not* be recorded unless considerations of forest productivity or biomass are more important than phytosociology."

9. Combining Mixed Data

It is rare in taxonomic studies to have attributes all of the same type. The date are usually a mixture with some binary, others meristic or continuous, and so on. Accordingly, it is imperative that some means of combining data from different attributes be available. Up to and including the publication of the book "Principles of Numerical Taxonomy" by Sokal and Sneath (1963) the majority of classificatory strategies were designed to operate on a single type of data and that usually settled on was binary (see Rogers and Fleming, 1964). An exception was the work of Sokal and Michener (1958) on bees and here they combined binary and meristic data into a correlation measure.

The manner in which disordered multistate data may be converted to binary has already been considered and the conversion of ordered multistate to binary is equally simple. Given the four grades "very long," "long," "short," and "very short," three binary attributes may be generated. The first of these is "long" versus "short" and the second and third of them are their consequential attributes "very long" versus "not very long" and "very short" versus "not very short."

To convert either meristic or continuous to binary data is straightforward but usually unsatisfactory, particularly if the frequency distribution of the attributes scores is approximately normal. Any distribution may be divided arbitrarily into two sections, thereby being converted to a binary attribute. With normal distributions the dividing line is usually the mean, in which circumstance two entities differing only slightly from one another but placed on either side of the mean become equally dissimilar to a pair drawn from the extremes of the range. Considered another way, all entities on either side of the mean acquire identical binary scores.

Combinations of meristic and continuous data are readily effected, for example, by rounding the latter to the nearest unit. Another way of handling the situation is exemplified by Hughes and Thomas (1971a) in a benthic study in Canada. Here the problem was somewhat different, with animals recorded meristically as numbers and plants recorded continuously as dry weights. The differences in the units and in the forms of data are both eliminated by standardizing to zero mean and unit variance. (See Chapter 7, Section D for discussions of standardization.) Likewise binary data may be made compatible with meristic or continuous data by a similar standardization. In these circumstances the resultant scores have only two values, the one corresponding to the binary score zero, the other to the score unity. Nonetheless these scores may be treated as any other standardized data.

The problems of handling mixed data and especially multistate data have been studied by Lance and Williams (1967b), who also considered the handling of matrices containing a relatively high proportion of missing or inapplicable entries. At that time four mixed data classificatory algorithms were known, due to Goodall (1966a), Lance and Williams (1967b), Gower (1967), and Burr (1968); Lance and Williams introduce a fifth based on heterogeneity as measured by an information statistic. Detailed consideration of this statistic will be deferred until later (see Appendix), but meanwhile it may be said that although the information statistic deals straightforwardly with discontinuous data it is less satisfactory with continuous. For convenience, such data are usually converted into ordered multistate attributes, and Lance and Williams (1967b) chose eight states (see Appendix). One should note that this analytical method incorporates joint absences as well as joint presences and is hence more suitable for taxonomic than for ecological classifications.

In ecological situations in which there is a small amount of binary data and the bulk is meristic, the easiest solution to the problem of combining the two kinds of data appears to be to count "presence" as equivalent to unity. In recent work Stephenson *et al.* (1974) effected this solution in dealing with grab catches of marine benthos. The animals were recorded in meristic form, but the marine plants were handled differently and recorded as presence or absence. For example, species of the marine plant *Halophila* reproduce by stolons and it is impossible to know how many individuals are contributing to a collection, though it may be possible to count the number of shoots.

6

Measures of Similarity and Difference

A. GENERAL

Similarity and difference are mutually dependent concepts and in much modern literature the former term applies to both. Nevertheless, the two will generally be distinguished in what follows when there is any possible ambiguity of meaning. It is clear that the concept of similarity has no meaning unless there are at least two things to be compared and it is equally clear that the apparent similarity of any two members of a population will depend on the total heterogeneity of that population. There are thus at least three concepts of similarity to be considered—that between entities, that between an entity and a group of entities, and that between groups of entities. There are further differences, for example, whether similarity between pairs of entities is determined with reference to or without reference to the population as a whole. The distinctions between these various concepts of similarity will become evident in the following discussion.

A wide variety of interentity similarity measures has been proposed but relatively few are in current use. The restriction in number has resulted from several causes. Many of the neglected indices are mere variants of others and have similar properties; some have unfavorable properties, especially as they approach their limits; others are highly specialized being based on a priori hypotheses concerning the nature of the populations being classified.

Some of the measures discussed below estimate dissimilarity rather than similarity but since the two are complementary concepts this need not cause any confusion. The reason for stressing dissimilarity in certain situations is that such measures are readily envisaged as "distances apart." This applies not only to Euclidean distance measures satisfying the theorem of Pythagoras but to many others as well, including some with the unusual property that as groups grow, the distances between them change in a nonlinear fashion.

Though most kinds of measures (of both similarity and dissimilarity) are interrelated and are readily interconverted they are, for convenience, discussed below under five separate headings each of which will now be considered briefly and then discussed at greater length. We appreciate that these headings omit some measures, for example; those based on variance.

1. Coefficients of Similarity

These coefficients are available for binary, graded, meristic, or continuous data. For binary data they are derived as a ratio involving the number of attributes shared by a pair of entities relative to the total number of attributes involved in the comparison. With most of these coefficients, values range from zero (nil similarity) to unity (complete similarity). If the similarity measure is designated as S its complement $1 - S$ is a dissimilarity measure.

2. Coefficients of Association

These measures may be estimated from binary, graded, meristic, or continuous data and reflect the manner in which the properties of pairs of entities are correlated. The magnitudes of these coefficients vary from $+1$ (indicating that for two variables a change in one is accompanied by an identical change in the other) to -1 (indicating that any change in one is accompanied by an equal and opposite change in the other). Such coefficients between entities indicate them to be similar or otherwise according as to whether the value of the coefficient approaches $+1$ or -1. If the similarity so estimated is designated as a then $(1 - a)/2$ may be regarded as an estimate of dissimilarity.

3. Euclidean Distance

The concept of Euclidean distance (D) as a dissimilarity measure has been introduced in Chapter 2, Section D and it applies equally also to meristic and continuous data. The Euclidean distance between entities may vary from zero (complete similarity) to an indefinitely large value depending on the number and magnitudes of the differences involved. Since Euclidean distance is always estimated via Pythagoras' theorem, D^2 is necessarily calculated and this too is often used as a dissimilarity measure. There is no direct similarity counterpart to this measure.

4. Information Content or Diversity Measures

Unlike the measures described above these are only incidentally interentity dissimilarity measures. They are basically measures of diversity within groups and are useful for generating classifications in that they enable groups to be united in such a way as to minimize the within-group diversity at each step. Since the first step involves the fusion of single entities, the increases in diversity following the fusions may be used as an interentity measure of dissimilarity. Here a zero increase in diversity (ΔI) following fusion of two entities indicates they are identical over the range of attributes considered. The greater the increase in diversity following fusion the less similar are the pair of entities. Hence, information or diversity measures like Euclidean distance may vary from zero to large values and have no similarity counterparts.

5. Similarity Measures Dependent on Probability Estimates

These measures tacitly assume the entities to be classified are samples from a larger population and that the probabilities of obtaining by chance pairs of entities as similar as those under consideration may be estimated. The probability estimate may be used as a dissimilarity measure and its complement as a similarity measure. Sometimes the probability is estimated via the prior determination of a χ^2 value whose role was to measure the degree of association between entities. In these circumstances the χ^2 value itself may be used as a dissimilarity measure, the greater its magnitude the less similar being the entities. Chapter 6, Section F and Chapter 8, Section C, 3).

B. COEFFICIENTS OF SIMILARITY

As stated above, similarity coefficients are closely related to dissimilarity coefficients, and many (but not all) range from zero at nil similarity to unity at complete similarity.

A great number of similarity coefficients or similarity indices is known. The best known have been listed and defined by Goodman and Kruskal (1954, 1959), Dagnelie (1966), Sokal and Sneath (1963), Macfadyen (1963), and Sneath and Sokal (1973). Of those considered below a few are primarily of historical interest but the majority are in current use.

It is pertinent to inquire into the circumstances leading to the development of such a variety of coefficients. Some reflect the need to accommodate particular forms of data, as for example, those restricted to binary data. Others allow for unevenness in the frequencies of attributes and minimize the influence of large or small values. Yet others are based on prior ideas concerning the statistical distributions of the properties measured.

In some instances each or all of these considerations may influence the choice of index. Accordingly, the definition of each of the indices that follow is in general accompanied by an account of its properties. Fortunately for a wide range of data most of the indices are monotonic* with respect to one another, but this observation should not be permitted to lull the user into believing then that all indices are interchangeable. Each stresses a particular property of the data and they do not all necessarily yield similar results when the entities whose similarities they measure are clustered.

1. Applying to Binary Data

To facilitate the comparison of the coefficients available for binary data a standard nomenclature will be adopted. It will be remembered that in normal taxonomic classifications the entities are taxa and the attributes their taxonomic features, and in ecological classifications the entities are sites and the attributes species. Consider a single attribute with, in either event, scores of 1 or 0. There are only four outcomes possible when comparing a pair of entities (taxa or sites). These are that both entities record the attribute in the first state (1, 1), that both record the alternative state (0, 0), or that one entity records one state and other records the alternative, i.e., (1, 0) or (0, 1). For a number of attributes we can obtain the summated values of each of the four possibilities. These values are given in parenthesis in the 2×2 table below.

		Taxon or site 2	
		1	0
Taxon or site 1	1	1, 1 (*a*)	1, 0 (*b*)
	0	0, 1 (*c*)	0, 0 (*d*)

* *Monotonic* in the present sense means *of similar sign*, i.e., when one increases so does the other. Nonmonotonicity occurs when an increase in one index accompanies a decrease in another.

Here the letters a, b, c, and d refer to the summated numbers of attributes, viz., a is (1, 1), the number of attributes in one state shared by both entities; b is the number of attributes for which the joint score is (1, 0), the number possessed by the first entity but not the second; c is the number of attributes for which the joint score is (0, 1), the number possessed by the second entity but not the first; and d is the number of attributes for which the joint score is (0, 0), the number possessed by both entities in the alternative state. The sum $a + b + c + d$ is the total number of attributes for which the entities have been compared.

As indicated earlier with taxonomic studies the score 0 often means the possession of an attribute state different from that scored as 1. For example, taxa may differ in flower color with one color scored as 1, the other as 0. In contrast, with ecological studies the scores 1 and 0 are usually referred to presence (1) or absence (0) so that the score (0, 0) indicates a conjoint absence and so d is the number of such absences.

Of the variety of similarity indices available and based on the above 2×2 tabulation, that which incorporates the unweighted totals from all cells is conceptually the simplest. Known as the Simple Matching Coefficient (SMC) it is defined as follows:

$$\text{SMC} = \frac{a + d}{a + b + c + d}$$

The SMC has no doubt been long known and it has been dissatisfaction with it that has generated many of the others. One source of this dissatisfaction has been the term d, when it refers to conjoint absences.

In some circumstances it would seem ridiculous to regard two taxa or sites as similar largely on the basis of them both lacking something. In other circumstances it would seem improper to neglect conjoint absences when estimating similarity.

In order to resolve these difficulties similarity coefficients with and without the inclusion of d have been designed and each group will now be considered. The former tend to be employed in taxonomic work and the latter in ecology.

a. Involving Conjoint Absences

In taxonomic work the status of d in similarity coefficients is ambiguous. In some instances it refers to conjoint occurrences of a given attribute state as with two plants possessing white rather than blue flowers; at other times it refers to the conjoint absence of a structure as with two beetles, each lacking elytra.

The following similarity coefficients incorporating d have been used in recent times and most assume values between zero for complete dissimilarity to unity for complete similarity. The range of values each assumes is given in parenthesis.

Simple matching:

$$\frac{a+d}{a+b+c+d} \quad \frac{\text{cooccurrences plus conjoint absences}}{\text{overall total of occurrences and conjoint absences}} \quad (0 \to 1)$$

Russell and Rao (1940)

$$\frac{a}{a+b+c+d} \quad \frac{\text{cooccurrences}}{\text{overall total of occurrences and conjoint absences}} \quad (0 \to 1)$$

Rogers and Tanimoto (1960)

$$\frac{a+d}{a+d+2(b+c)}$$ as simple matching except mismatches carry double weight $(0 \to 1)$

Hamann (1961)

$$\frac{(a+d)-(b+c)}{a+b+c+d}$$ as simple matching except numerator reduced by mismatches $(-1 \to +1)$

Sokal and Sneath (1963)

$$\frac{2(a+d)}{2(a+d)+b+c}$$ as simple matching except cooccurrences and conjoint absences carry double weight $(0 \to 1)$

b. Neglecting Conjoint Absences

Such indices are particularly useful in ecological studies when a series of sites are being compared and these possess a few species common to each with the remainder restricted to few of the sites.

Most of the coefficients incorporating d are constrained between zero and unity. In the present measures which eliminate d, low similarities may be either positive or negative and near to zero, while high similarities assume values from near to unity to infinity.

Jaccard (1908)

$$\frac{a}{a+b+c} \quad \frac{\text{cooccurrences}}{\text{total occurrences at either site}} \quad (0 \to 1)$$

Czekanowski (1913)

$$\frac{2a}{2\ \ + b + c}$$ as for Jaccard except coincidences carry double weight $(0 \rightarrow 1)$

Kulczynski (1927)
first

$$\frac{a}{b + c} \quad \frac{\text{cooccurrences}}{\text{mismatches}} \quad (0 \rightarrow \infty)$$

Kulczynski (1927)
second

$$\frac{a}{2}\left(\frac{1}{a + b} + \frac{1}{a + c}\right) \quad (0 \rightarrow 1)$$

Ochiai (1957)

$$\frac{a}{\sqrt{(a + b)(a + c)}} \quad (0 \rightarrow 1)$$

Fager and McGowan (1963)

$$\frac{a}{\sqrt{(a + b)(a + c)}} - \frac{1}{2\sqrt{a + b}} \quad (< 0 \rightarrow < 1)$$

In choosing a coefficient, it would seem desirable to avoid those not constrained between 0 and 1 unless their use can be justified. The index quoted which tends to infinity is clearly sensitive to small changes (particularly in a) and this situation could well arise from repeated sampling of the same community in certain ecological situations.

Possibly the Jaccard and Czekanowski coefficients are the most satisfactory of those in widespread use, for example, the latter has been employed in marine ecological studies by Stephenson *et al.* (1970) (where it was mistakenly called the Jaccard) and by Gage (1972) where it was referred to its better known later author, Sørensen (1948). Field and his co-workers in South Africa frequently refer to the Czekanowski coefficient but mostly use its equivalent for meristic and continuous data the Bray-Curtis measure.

To assist familiarity with the coefficients generally and to illustrate the differences between the Jaccard and Czekanowski, and example is given in Table 6.1.

The presence and absence of 10 attributes from four entities are listed, and coefficients are given between entities A and B and between C and D.

TABLE 6.1

THE JACCARD AND CZEKANOWSKI COEFFICIENTS FOR TWO PAIRS OF ENTITIES A–B AND C–D

	Entities			
Attributes	A	B	C	D
1	1	1	1	0
2	1	0	0	0
3	0	1	0	1
4	1	1	1	1
5	0	0	0	1
6	1	0	0	0
7	0	1	0	0
8	0	0	0	1
9	1	1	1	0
10	1	0	0	0
	$a = 3$		$a = 1$	
	$b = 3$		$b = 2$	
	$c = 2$		$c = 3$	
	$d = 2$		$d = 4$	
	A–B		C–D	
	Jaccard 3/8 = 0.38		1/6 = 0.17	
	Czekanowski 6/11 = 0.55		2/7 = 0.29	

In both cases there are relatively few conjoint presences (especially in the second), and the larger values of the Czekanowski coefficient are likely to be the more attractive. It is easy to show that with relatively many conjoint presences the Jaccard coefficient will be more attractive because it will give a wider spread of values near the upper end of the range.

The index of Fager and McGowan merits special attention because it has been associated with a particular form of nonhierarchical inverse "clumping"—the recognition of "recurrent species groups," and these have been widely used in marine ecology. Examples include Fager and McGowan (1963), Sheard (1965), Fager and Longhurst (1968), Fager (1968), Jones (1969), Longhurst (1969), Lie and Kelley (1970), Bayer *et al.* (1970), Martin *et al.* (1970), Boesch (1971), and Bowman (1971).

There are general objections to nonhierarchical methods which we have considered elsewhere (Chapter 4, Section B) and some specific objections to the Fager and McGowan index as outlined by McConnaughey (1964) and Field (1971). Further objections are, first, that similarity may approach but never attain the value of unity, second, that when a is small compared with b or c similarity may have a small negative value, and, third, that the index is assymetric in that the comparison of entity A with entity B yields different values from the comparison of entity B with entity A.

The Fager and McGowan index is derived from Ochiai's (1957) which appears to have been little used and seems attractive. Its denominator involves a geometric mean which in some circumstances is likely to be a more effective "standardization" than a sum. The introduction of a "correction factor" by Fager and McGowan does not appear, in our view, to have improved the original index.

We have not given in the main list one of the more complex of the coefficients used for binary ecological data and which neglects conjoint absences. This is derived from Preston's (1962) resemblance equation, which in the present notation is

$$\frac{(a+c)^n}{a+b+c} + \frac{(a+b)^n}{a+b+c} = 1$$

where n is the resemblance. The equation is transcendental with no general solution, however a set of approximate solutions is given in Preston's publication (1962, p. 419). Preston prefers to use the reciprocal of n which he denotes as z and which has the advantage of being constrained between zero for similarity and unity for dissimilarity.

Preston's measure of resemblance is based on the assumption that there is a lognormal distribution of individuals in natural species assemblages. The measure was designed primarily for biogeographical work, but has been used in phytoplankton work by Thorrington-Smith (1971). It seems doubtful whether a coefficient of resemblance based on these premises and of this order of complexity is justifiable in dealing with binary data.

2. Applying to Meristic and Continuous Data

Again many coefficients are available; we shall only consider those we have used or noted in recent literature.

Two dissimilarity measures are in reasonably common recent use. The first is a dissimilarity measure widely known as the "Bray-Curtis" measure. It is the complement of that used by Bray and Curtis (1957) with a further difference that Bray and Curtis used it for standardized data. We refer to it subsequently as the Bray-Curtis measure, without quotes. It was actually used by earlier workers (Motyka *et al.*, 1950) and is a quantitative extension of the complement of Czekanowski's (1913) coefficient. Its use is easiest to illustrate in an ecological context. If n is the number of attributes (species) and x_{1_j} and x_{2_j} are the values of the jth attribute for any pair of entities (sites) then the coefficient is

$$\frac{\sum_{1}^{n} | x_{1_j} - x_{2_j} |}{\sum_{1}^{n} (x_{1_j} + x_{2_j})}$$

In this expression $| x_{1_j} - x_{2_j} |$ indicates that the value of the difference is always positive. It should be noted that in this coefficient the denominator is a sum involving all individuals of all species at the the two sites; hence, it tends to be greatly influenced by occasional outstanding values. This coefficient has been used in recent marine studies by Field and Macfarlane (1968), Field (1969, 1970, 1971), Day *et al.* (1971), Stephenson and Williams (1971), and Stephenson *et al.* (1972).

Another dissimilarity measure was first used by Lance and Williams (1966a) and termed the "non-metric" coefficient. Later its metric properties were established, and it is now known as the "Canberra metric" (Lance and Williams, 1967b). Using the previous notation it is

$$\frac{1}{n} \sum_{1}^{n} \frac{| x_{1_j} - x_{2_j} |}{(x_{1_j} + x_{2_j})}$$

It differs from the Bray-Curtis in being the sum of a series of fractions and, hence, an outstandingly abundant species or an outstanding difference can only contribute to one of the fractions and so does not come to dominate the index.

The incorporation of zero values into the Canberra metric ls subject to certain conventions. When one element of any comparison is zero, the term:

$$\frac{| x_{1_j} - x_{2_j} |}{(x_{1_j} + x_{2_j})}$$

becomes unity. This applies no matter how large or small the other element. Thus using the Canberra metric elements, a pair of elements with values of 1000 and 0, respectively, will appear equally dissimilar to a pair with values of 0.1 and 0.0. This would appear to be unreasonable and the problem may be circumvented by replacing those zero values, which enter into one of the above terms, by a small positive number. Choice of the number is arbitrary but it should be smaller than any of the recorded values in the population under study and its efficiency may be judged by trial and error. Stephenson *et al.* (1972) used a value one-fifth that of the lowest entry in their data matrices.

When both x_{1_j} and x_{2_j} are zero, their difference over their sum is taken to be zero instead of indeterminate. Hence such comparisons add nothing to the summation part of the index, but nevertheless cause a decrease in magnitude since the number of comparisons is enlarged in the divisor. Thus comparisons of sites with few species in common will lead to small values of the Canberra metric suggesting the sites to be less dissimilar. The apparent similarity is merely a reflection of their both lacking most of the species present in the sample of sites being studied.

If it is desired that double zero records be ignored to avoid this, then the appropriate divisor is not n but $n - r$ where r is the number of comparisons involving double zero comparisons.

Confusion may sometimes arise in the minds of potential users over the difference between the Bray-Curtis and Canberra metric measures of dissimilarity. This may be resolved by consideration of the following example in which the site entries refer to the numbers of each species present.

	Sites	
Species	A	B
1	10	5
2	5	1
3	0	1
4	1	0
5	0	0
6	10,000	10

Bray-Curtis dissimilarity measure

$$= \frac{(10-5)+(5-1)+(1-0)+(1-0)+(10{,}000-10)}{(10+5)+(5+1)+(1+0)+(1+0)+(10{,}000+10)}$$

$$= \frac{5+4+1+1+9990}{15+6+1+1+10{,}010}$$

$$= \frac{10{,}001}{10{,}033}$$

$$= \text{ca. } 1.00$$

Canberra metric dissimilarity

$$= \frac{1}{6}\left(\frac{5}{15} + \frac{4}{6} + \frac{1}{1} + \frac{1}{1} + \frac{9990}{10010}\right)$$

$$= \frac{0.33 + 0.67 + 1 + 1 + \text{ca. } 1}{6}$$

$$= \text{ca. } 0.66$$

The dominant influence of the large value of 10,000 in determining the magnitude of the Bray-Curtis measure is very apparent here.

It will be noted that both the Bray-Curtis and the Canberra metric measures of dissimilarity involve at each stage only the pair of entities under consideration. The properties of the total population of entities are not involved.

A number of other dissimilarity measures have been proposed in which the population parameters are also involved and these will now be discussed in chronological order of publication together with some of their properties.

When describing two new subspecies of leaf-nosed snake Klauber (1940) proposed a coefficient of divergence (C.D.) defined as follows:

$$\text{C.D.} = 2\,\frac{(\bar{m}_1 - \bar{m}_2)}{\bar{m}_1 + \bar{m}_2}$$

where $\bar{m}_1$ and $\bar{m}_2$ are the mean values of a parameter in two populations. Though not stated it may be assumed the smaller is always subtracted from the larger mean so as to keep the coefficient positive.

Though the coefficient is of limited value referring as it does to a single attribute, and applying to pairs of populations rather than pairs of entities, it is included here because of its evident relationship to the Canberra metric.

The measure proposed by Cain and Harrison (1958) also depends on a knowledge of the whole set of entities being classified. It was named the mean character difference (M.C.D.) between entities and is defined as:

$$\text{M.C.D.} = \frac{1}{n}\sum_{1}^{n}\frac{|\,x_{1_j} - x_{2_j}\,|}{x_{\max}}$$

where x_{1_j} and x_{2_j} are the values of the jth attribute for every pair of taxa and $x_{\max}$ is the maximum value assumed by the attribute. Hence the value of the index depends at every step on a knowledge of the total population as well as the pair under consideration. The values of the coefficient vary from zero (complete similarity) to values approaching but never equaling unity (complete dissimilarity) unless the smallest score for an attribute is zero. In both these respects the M.C.D. resembles the Canberra metric but unlike that measure does not give constant similarity values when magnitudes are different but proportions are unchanged.

A final measure of dissimilarity attributed to Gower by Sheals (1964) and used by a number of other taxonomists (Jamieson, 1968; Sims, 1966) is of the form:

$$\frac{1}{n}\sum_{1}^{n}\frac{|x_{1_j}-x_{2_j}|}{\text{range}}$$

As with the Bray-Curtis and Canberra measures this coefficient assumes values from zero (similarity) to unity (dissimilarity).

C. COEFFICIENTS OF ASSOCIATION

1. General

A series of coefficients has long been available for estimating the extent of association between attributes within populations. If, for example, it were desired to determine whether there is any association between sexuality (male or female) and handedness (left or right) then this could be checked by counting the numbers of right- and left-handed males and females in a sample or population and expressing the result in terms of one of the coefficients of association given below. The coefficients may also be employed as similarity measures, the sites or taxa now being compared in pairs with respect to the set of attributes available.

While tests of significance are available for most of these coefficients they are in general appropriate only when the association between attributes is being estimated. In these circumstances each entity (site or taxon) is independent of the others, and further with large populations a series of entities may be chosen at random. In contrast, when the coefficients are employed as similarity measures between entities there is no guarantee the attributes are independent. Furthermore, as it is usual for all the attributes and not just a random sample to be employed in estimating the magnitude of the coefficient, tests of statistical significance are then inappropriate.

Most of the coefficients given below are constrained between +1 and −1. Thus two taxa or sites with a coefficient +1 would be similar in all respects, whereas if the coefficient were −1 they would be dissimilar in all respects. Values of the coefficient of zero indicate that the pair of entities under consideration may be regarded as equally similar or equally dissimilar.

There seems to be little advantage in using these coefficients when similarity measures constrained between zero and unity are available, which fact may account for their general neglect in numerical classificatory studies.

The two groups of coefficients of association will now be considered in some detail.

2. Applying to Binary Data

a. Constrained between +1 and −1

Of the three measures to be considered two take into account conjoint absences but the third, due to McConnaughy (1964), neglects such conjoint absences. In terms of the variables as defined above for coefficients of similarity the coefficients of association are

Yule

$$\frac{ad - bc}{ad + bc}$$

Pearson

$$\frac{(ad - bc)}{(a + b)(c + d)(a + c)(b + d)}$$

McConnaughy

$$\frac{a^2 - bc}{(a + b)(a + c)}$$

The above coefficients of association, though not widely used, have been employed in a number of numerical taxonomic and ecological studies. The McConnaughy coefficient, designed originally for plankton work, was employed by Field (1971) as an association measure for an inverse analysis of South African benthos. In their study of marine benthos Lie and Kelley (1970) used the Pearson index as an association measure.

As indicated earlier the coefficients of similarity and dissimilarity constrained between zero and unity are interrelated as the complement of each other. Though they appear not to have been used in the form that follows, the three association coefficients just described are readily converted to dissimilarity (and with a slight modification similarity) coefficients by means of the following relationship.

$$\text{Dissimilarity} = \frac{(1 - \text{coefficient of association})}{2}$$

Thus considering the Yule coefficient as an example

$$\text{Dissimilarity} = \frac{bc}{ad + bc}$$

When the entities are alike in all respects b and c are zero and so the dissimilarity is zero; when the entities are alike in all respects a and d are zero and so the dissimilarity is unity, its maximum value.

b. VALUES RANGING FROM ZERO TO INFINITY

The second group of measures with no negative values and no upper limit are the chi-square (χ^2) and its scaled variant the mean square contingency (M.S.C.). These are defined as follows:

$$\chi^2 = \frac{(ad - bc)^2 \times (a + b + c + d)}{(a + b)(c + d)(a + c)(b + d)}$$

and

$$\text{M.S.C.} = \frac{\chi^2}{a + b + c + d}$$

Where there are no missing data as with the majority of presence–absence studies of species in sites χ^2 does not need to be scaled by dividing by the number of comparisons considered. In contrast, comparisons of pairs of taxa often involve comparing differing numbers of attributes due to some being either missing or consequential and so scaling is required as with the M.S.C.

Neither of these indices appear to have been used in classificatory programs involving clustering but the χ^2 in particular has been used widely as a measure of association in programs involving fission of groups (see Chapter 8, Section C, 3).

3. Applying to Meristic and Continuous Data

The measure of association most commonly used for meristic and continuous data is the product–moment correlation coefficient (r), hereinafter referred to as the correlation coefficient. It was designed to measure the association between pairs of variables such as height and weight, there being a separate height and weight for each entity. The measure as originally defined is therefore an estimate of the correlation between a given pair of attributes. If plotted against Cartesian axes each point of the resultant scatter diagram would refer to the same pair of attributes and would be an expression of the magnitude of those attributes for a given entity.

A simple symbolic expression of the product–moment correlation coefficient is

$$r = \frac{\sum_{1}^{n} (x - \bar{x})(y - \bar{y})}{\sqrt{\sum_{1}^{n} (x - \bar{x})^2 \sum_{1}^{n} (y - \bar{y})^2}}$$

where n is the number of entitities, $\bar{x}$, $\bar{y}$ are the mean values of the attributes in the entities and x and y are the individual measurements of a given pair of attributes.

The correlation coefficient has been employed with considerable success as a similarity measure, especially in those situations where absolute size alone is regarded as less important than shape. Thus in classifying plants and animals, the absolute sizes of the organisms or their parts, within limits, are often of less impoıtance than their shapes. And so where it is suspected that the differences in sizes observed between organisms are merely a reflection of their different ages or a reflection of their responses to a nonuniform environment, it is important that the influence of such differences be minimized and, in general, the use of the correlation coefficient as a similarity measure will achieve this purpose.

Taxonomic examples of the use of r as a similarity measure are due to Sokal and Michener (1958) who used it for studies on bees, Rhodes *et al.* (1970, 1971) to measure similarities between mango and avocado species, respectively, and Boyce (1969) to measure similarities between apes and their relatives.

The use of r as a similarity measure in normal analyses requires that the taxa or sites be compared in pairs over the range of attributes measured. Here each one of a pair of Cartesian axes represents an entity and each point of the scatter diagram resulting from comparing the entities indicates the magnitude of a separate attribute. Thus one point might reflect the leaf size in the two entities, another point the fruit weight, and so on over the whole set of attributes.

In ecological studies the use of r as a similarity measure has been restricted largely to species/species classifications where its ability to distinguish nil correlations from negative correlations gives it an advantage over some other coefficients of association. With site/site classifications it is usual to find that many of the r values estimated are small due to a few species being abundant and the others rare. For this reason, although r has been used extensively in marine studies, for example, by Cassie (1961) and Cassie and Michael (1968), its use requires caution. Field (1970) has proposed that when 50% or more of the data matrix entries are zero, the correlation coefficient as a measure of site similarity should be avoided. In benthic surveys this is only possible by drastic elimination of the less common species.

While it has been recommended that r be used as a similarity measure when similarity in shape rather than size is required, it should be noted that in the limit when two entities are identical in shape and differ only in size r may be indeterminate. For example, if the similarity between two cubes with sides say 4 and 6 units, respectively, is measured in terms of the correlation coefficient both the numerator and denominator employed in this estimation are zero and will likewise be zero for all comparisons between cubes. The existence of indeterminate values of r under these circumstances indicates that the role of r as a similarity measure requires further investigation.

D. EUCLIDEAN DISTANCE AS A DISSIMILARITY MEASURE

We have already introduced this concept (see Chapter 2, Section D and Chapter 6, Section A) as taxonomic distance and ecological distance is its exact equivalent. In essence it is the distance between two entities whose positions are determined with respect to their coordinates, these being defined with reference to a set of Cartesian axes (i.e., axes at right angles to each other). It is a dissimilarity measure which can be applied to both binary and continuous data.

The binary situation has already been outlined and we merely add that to avoid the calculation of square roots Euclidean distance (D) is often replaced with Euclidean distance squared (D^2). When using binary data replacement of D by D^2 has the further advantage that the difference between entities measured in the latter scale is nothing more than the number of attributes by which they differ, provided there are no missing data so all interentity values are based on the same number of comparisons. In the notation already developed Euclidean distance squared (D^2) is $b + c$ and the Euclidean distance is $\sqrt{b + c}$.

This may be readily seen if we regard the binary data 0, 1 as scored to be represented on Cartesian axes. For a pair of axes and attributes the four possible coordinates would then occupy positions as indicated in Diagram 6.1.

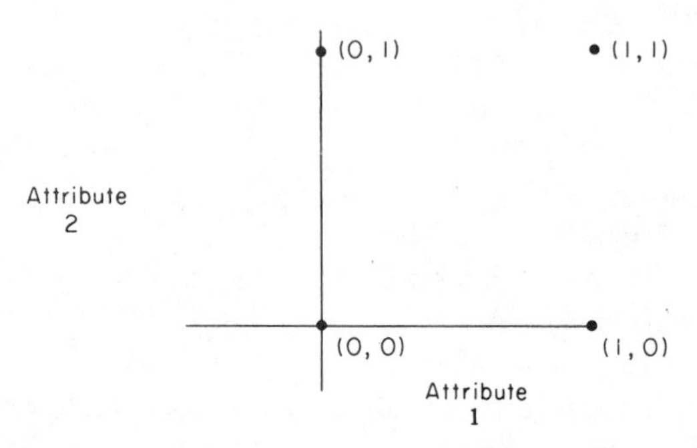

Thus two entities differing with respect to attribute 2 but not attribute 1 would be a single unit apart; two entities differing with respect to attribute 1 but not attribute 2 would likewise be a single unit apart; two entities differing with respect to both attributes would now be $\sqrt{2}$ units apart.

The addition of further axes produces a hypercube with sides of unit length. Hence b and c are the numbers of attributes with the scores dissimilar in the two entities and so represent the number of sides of the hypercube to be considered in calculating the diagonal. This distance is $\sqrt{b(1)^2 + c(1)^2}$ and not $\sqrt{b^2 + c^2}$ as might have been expected because of the unit lengths of the side.

With continuous data and with respect to any given attribute the Euclidean distance (D) between two entities is $| x_1 - x_2 |$ where x_1 is the score for one entity and x_2 that for the other. For n attributes

$$D = \left[\sum_1^n (x_1 - x_2)^2 \right]^{1/2}$$

where x_1 and x_2 are successively the scores for the n attributes.

To ensure that attribute scores are additive it is common practice to use D^2 rather than D as a measure of similarity. The manner in which such additivity is achieved will be illustrated with respect to the following data. Consider two entities differing with respect to three attributes as in Table 6.2.

If single attributes only are involved in the estimation of difference, D is more readily determined than D^2. If however, two or more attributes are considered D^2 is the easier to determine. For example, D^2 based on at-

TABLE 6.2

EUCLIDEAN DISTANCE (D) AND ITS SQUARE (D^2) FOR A PAIR OF ENTITIES A AND B

	Entity		D	D^2
Attribute	A	B		
1	3	5	2	4
2	4	8	4	16
3	5	14	9	81

tributes 1 and 2 has the value $2^2 + 4^2 = 20$, whereas the calculation D involves the further determination of a square root. Here $D = \sqrt{20} = 4.47$. Still more important is that the contributions of further attributes build more simply on to any already determined value of D^2 than D. Thus the incorporation of the contribution of attribute 3 to D^2 merely involves a further addition. That is D^2 now becomes $20 + 9^2 = 101$. In contrast the contribution of attribute 3 to the predetermined value of D based on two attributes demands that before the new information can be added, the value of D must be squared. That is D would now be determined as $\sqrt{(4.47)^2 + 9^2} = \sqrt{20 + 81.00} = \sqrt{101} \simeq 10$. The use of D^2 rather than D is therefore merited both on the grounds of additivity and ease of calculation.

As with the binary situation when the comparisons between entities involve different numbers of attributes the scores for either D or D^2 should be scaled to a common denominator by dividing by n, the number of comparisons available.

As a dissimilarity measure D^2 has the important and sometimes undesirable property that single large differences will come to dominate several smaller differences. In an ecological context it gives considerable weighting, possibly undue weighting, to the abundant species. For this reason the calculation of D^2 is frequently preceded by a transformation of the original data (see Chapter 7, Section C).

E. INFORMATION THEORY MEASURES OF SIMILARITY/DISSIMILARITY

Information theory finds a wide range of applications in classificatory problems due to its ability to measure disorder. The word information is used here in a technical sense which differs somewhat from common usage.

At first the terms "disorder" or "diversity" and "information" may appear unrelated but historically they are not, in that information theory grew out of the study of message transmission, where the number of symbols in the code reflected its diversity and the relative frequencies of the symbols reflected their information content.

However, the distinction between diversity and information has changed with usage. In a later section different concepts of diversity measures of sites are discussed. We can now note that originally diversity was merely the number of species present and paralleled the usage given above, namely, the number of symbols in the code. In its current use in both ecology and information theory, diversity involves relative frequencies (of species and symbols, respectively) and when using information theory measures, diversity and information content have come to mean the same thing.

The basic measure of diversity by information content is the Shannon diversity index H.* In the context of a site diversity this is

$$H = N \log N - \sum_{1}^{S} n \log n \quad \cdots \qquad (6.1)$$

where N is the grand total of individuals of all species at the site, S the number of species, and n the number of individuals in a given species. The derivation of this formula is given in Appendix.

The magnitude of the diversity depends upon two things. The first is the base of the logarithms which are used. Logarithms to three different bases e, 2, and 10 are in common use and tables of $n \log n$ to each of the bases are available. Herein values to the bases e and 10 have been employed almost exclusively and the base is cited if ambiguity might otherwise result. To convert from base e to base 10 results should be multiplied by 0.4343 and to convert from base 10 to base e they should be multiplied by 2.3026.

The magnitude of diversity also depends on the number of individuals collected, and hence increases with sampling intensity. For this reason it is sometimes desirable to express the diversity of a given site in terms of "per individual specimen." The formulation then becomes

$$H_{(1)} = \log N - \frac{1}{N} \sum_{1}^{S} n \log n \quad \cdots \qquad (6.2)$$

Examples may assist familiarization. Suppose a given site (site 1) has 5 individuals of species A, 3 of species B, and 2 of species C, and 0 of species

* We use H for Shannon diversity (not H') and use $H(B)$ later for Brillouin diversity.

D. From formula 6.1 the Shannon diversity is given by

$$H = 10 \log 10 - (5 \log 5 + 3 \log 3 + 2 \log 2 + 0 \log 0)$$

We neglect the term 0 log 0 which is zero times minus infinity and get $H = 4.47$. Per individual specimen by formula 6.2 this becomes

$$H_{(1)} = \frac{4.47}{10} = 0.447$$

In the above we have been considering site diversities in terms of the numbers of individuals of different species in a given site. It is equally open to us to consider species diversities in terms of the recordings of a given species in a number of sites. Formulas 6.1 and 6.2 are the same, the former being

$$H = N \log N - \sum_{1}^{S} n \log n$$

However the symbols now have different meanings. N is the grand total of the recordings of the species in all sites, S is the number of sites, and n is the number of individuals of the species at a given site.

We can now consider diversity values in a site/species matrix, and shall use the example below with original data within the body of the table.

The diversities of individual species are based on the data in the rows and the diversities of individual sites on the data in the columns. Next we consider the diversities of pairs of sites. For example, the species recordings

TABLE 6.3

THE INDIVIDUAL SPECIES AND SITE DIVERSITIES FOR A SAMPLE OF 3 SITES WHICH INCLUDE 4 SPECIES[a]

	Sites			Individual species diversities
Species	1	2	3	
A	5	7	0	3.54
B	3	2	7	4.79
C	2	1	2	2.29
D	0	0	1	0
Individual site diversities	4.47	3.48	3.48	

[a] See text.

at sites 1 and 2 are A-12, B-5, C-3, D-0 and the diversity (H) of the combined pair of sites is

$$H = 20 \log 20 - (12 \log 12 + 5 \log 5 + 3 \log 3) = 8.14$$

Comparable values for sites 2 and 3 and 1 and 3 are 15.93 and 10.12, respectively.

These values of H may be employed as a dissimilarity measure by determining the increase in diversity which has taken place as a consequence of pooling the data from the two sites. In algebraic form this may be expressed as

$$\Delta I = IC - (IA + IB)$$

Where ΔI is the increase in diversity, IC is the diversity of the combined sites and IA and IB are the diversities of the individual sites before fusion. The change of symbolism from H to I is unfortunate but is in line with current practice.

$$\begin{aligned} \text{For sites 1 and 2 } \Delta I &= 0.19 \\ \text{2 and 3 } \Delta I &= 8.97 \\ \text{1 and 3 } \Delta I &= 2.17 \end{aligned}$$

It should be noted that when the sites are identical with respect to species and their abundances, ΔI is zero, as can be determined by reference to the following example. Consider 2 sites each with 3 species as in Table 6.4.

TABLE 6.4

DIVERSITIES OF TWO SITES WITH IDENTICAL SPECIES ATTRIBUTES

	Sites		
Species	1	2	1 + 2
A	5	5	10
B	3	3	6
C	2	2	4
	Individual site diversities		Combined site diversity
	4.47	4.47	8.92

It is clear that here

$$\Delta I = IC - (IA + IB)$$

$$= 8.92 - (4.47 + 4.47) = 0$$

Several properties of the information measure and its derived dissimilarity should be appreciated. The limits of I ($=H$) range from negative (when N is less than unity) to infinitely large. A zero value of the Shannon index means only a single attribute is present since then $N \log N$ is numerically equal to $n \log n$. When two sites are fused, even if they are identical, as in the example above, the diversity of the resultant pair is greater than that of either considered singly. Nonetheless there is no increase in diversity over the sum of the diversities of the individual sites.

When the pair of sites are not identical and each site contains at least one species, on fusion there is always an increase in diversity over and above the sum of the two initial diversities. This property of diversity per site to increase with the size of the population recorded, whether it be a consequence of pooling quadrat data or arises from an increase in quadrat size should be carefully noted.

In contrast the diversity per individual is independent of sample size and is the same for sites in which the proportions of the individuals present do not differ. Thus, in the last example worked out each site has 10 individuals and so the diversity per individual is one-tenth that of the diversity per site; the combined sites contain 20 individuals and so the diversity per individual is one-twentieth that of the diversity per combined site. In both cases the diversity per individual is 0.447.

It will be evident that values can be obtained for ΔI when comparing species diversities between instead of within sites and these can be used as dissimilarity measures for inverse analyses.

It will be noted that information gain ΔI may be used as a measure of dissimilarity between entities, between entities and groups, and between groups themselves. The main value of information gain is less as a measure of dissimilarity and more as a clustering strategy in which there is a successive fusion into groups in such a manner that ΔI is minimal at each fusion. We shall consider information measures later under clustering strategies.

An information measure that is a direct similarity coefficient has been proposed by Hawkesworth *et al.* (1968) who employed it for studying similarities among attributes. The index S.I. is defined as follows:

$$\text{S.I.} = \sqrt{1 - (I_{AB})^2}$$

where I_{AB} is the interdependence between attributes measured as:

$$I_{AB} = \frac{\text{information held exclusively by (A) plus information held exclusively by (B)}}{\text{total information possessed by (A) + (B)}}$$

The properties of these measures are further discussed in the Appendix.

The Shannon diversity measure assumes that we are dealing with samples from an infinite population, i.e., one not affected by the sampling. If instead the samples are accepted as total population, as frequently they must be, it is more appropriate to employ the Brillouin diversity measure (Brillouin, 1962). The relative merits of the two measures have been discussed in some detail by Pielou (1966, 1969) and their derivation is given in the Appendix of the present work. Here it will suffice to say that for most classificatory purposes the employment of one rather than the other will have little affect, for the ratio of the two is almost constant over a wide range of n.

The Brillouin measure of diversity H (B) is defined as follows:

$$H(B) = \log N! - \sum_{1}^{S} \log n! \cdots \tag{6.3}$$

where, as before, N is the total number of individuals concerned, S is the number of species, and n is the number of individuals in a given species. As with the Shannon measure the Brillouin measure may be used to measure similarity between pairs of sites or species by subtracting from their joint diversity the summated diversities of the individual sites or species.

The Brillouin measure is being increasingly employed in preference to the Shannon, for example by Williams (1972) in partitioning information into its qualitative and quantitative components, and by Pielou (1972) for quantification of the concepts of habitat width and overlap by a more standard form of partitioning. This we shall refer to in Chapter 12, Section G.

We shall now consider information theory measures as applied to binary data. The obvious method for ecological data is to regard "presence" as a unit recording and follow previous procedures. For example, the data used in Table 6.3 become Table 6.5.

Comparing diversity values with those in Table 6.3 indicates immediately the loss of information about the sites when data lose their meristic form, for the abundances of the species are no longer available. In the same manner as previously we can calculate diversities of fusions of sites and

TABLE 6.5

TABLE 6.3 DATA CONVERTED TO BINARY FORM

Species	Sites 1	Sites 2	Sites 3	Individual species diversities
A	1	1	0	2 log 2 = 0.60
B	1	1	1	3 log 3 = 1.43
C	1	1	1	3 log 3 = 1.43
D	0	0	1	1 log 1 = 0
Individual site diversities	3 log 3 = 1.43	3 log 3 = 1.43	3 log 3 = 1.43	

thus ΔI. For example,

$$I_{1+2} = 6 \log 6 - 3(2 \log 2) = 4.67 - 1.806 = 2.86$$

$$\Delta I_{1+2} = 2.86 - 2(1.43) = 0$$

$$I_{1+3} = 6 \log 6 - (1 \log 1 + 2 \log 2 + 2 \log 2 + 1 \log 1) = 4.67 - 1.20$$
$$= 3.47$$

$$\Delta I_{1+3} = 3.47 - 2(1.43) = 0.61$$

$$\Delta I_{2+3} = \Delta I_{1+3} = 0.61$$

Though it is clearly possible to treat binary data in terms of the above information measures, this does not appear to have been followed. One reason has been given, to do so having collected meristic data would mean discarding most of the information collected. Such a move would not be worthwhile unless it was for a special reason such as enabling data to be pooled where some was in the binary and the remainder in the meristic state. There is, however, another reason and this stems from the application of information measures to taxonomic situations.

Here the binary data often contrast pairs of positive states, for example, fruit dehiscent or indehiscent. In this situation two plants with dehiscent fruits are quite as alike to one another as are two plants with indehiscent fruits. Likewise in ecological studies it may be desirable to regard two sites as alike if they both lack a given species but there are few times when such a viewpoint is adopted.

In its usage as developed below, information theory does not determine the diversity of the entities individually and then after fusion but proceeds directly to measure ΔI, the increase when fusing pairs of entities, pairs of groups of entities or individuals to existing groups. The determination of ΔI is given in the formula below and its derivation is given in the Appendix.

$$\Delta I = SN \log N - \sum_{1}^{S} [a_j \log a_j + (n - a_j) \log (n - a_j)]$$

where N is the total number of entities in the two groups, S is the number of binary attributes possessed by the set of entities involved, and a_j is the number of species possessing one of the binary states of each attribute and $N - a_j$ is clearly the number of species possessing the alternate binary state.

In order to illustrate the application of ΔI in this sense reference will be made again to Table 6.5 and both the normal and inverse analyses will be undertaken. Since the data matrix has the differing number of rows and columns, the number of entities (N) will be 3 in the normal analysis and 4 in the inverse analysis. The two states of the binary attributes 1 and 0 will be defined so that a_j is the number of entities in state 0; the number of attributes S will be 4 in normal analyses and 3 in the inverse analysis.

Considering the fusion of sites in pairs the following results are obtained:

$$\Delta I = S(N \log N) - \sum_{1}^{S} [a_j \log a_j + (N - a_j) \log (N - a_j)]$$

$$\Delta I_{1+2} = 4(2 \log 2) - \begin{bmatrix} 0 \log 0 + 2 \log 2 \\ 0 \log 0 + 2 \log 2 \\ 0 \log 0 + 2 \log 2 \\ 2 \log 2 + 0 \log 0 \end{bmatrix}$$

$$= 4(2 \log 2) - 4(2 \log 2) = 0$$

$$\Delta I_{1+3} = 4(2 \log 2) - \begin{bmatrix} 1 \log 1 + 1 \log 1 \\ 0 \log 0 + 2 \log 2 \\ 0 \log 0 + 2 \log 2 \\ 1 \log 1 + 1 \log 1 \end{bmatrix}$$

$$= 4(2 \log 2) - 2(2 \log 2) = 2(2 \log 2)$$

Likewise

$$\Delta I_{2+3} = 2(2 \log 2)$$

Thus as expected, no increase in diversity follows the fusion of sites 1 and 2, since these are identical in species composition. The fusion of sites 2 and 3 is in each instance accompanied by increase of diversity equal to 2(2 log 2). The exact value of this expression will, of course, depend on the base of the logarithms employed. To the base 2, 2(2 log 2) = 4.00; to base e = 3.77, to base 10 = 1.20.

The inverse analysis in which the species are compared in terms of their site presence or otherwise (i.e., sites are attributes) gives rise to the following values of ΔI.

$$\Delta I_{A+B} = 3(2 \log 2) - 2(2 \log 2) = 2 \log 2 = \Delta I_{A+C}$$

$$\Delta I_{B+C} = 3(2 \log 2) - 3(2 \log 2) = 0$$

$$\Delta I_{A+D} = 3(2 \log 2) - 0.0 = 3(2 \log 2)$$

It will be noted that the expression $a_j \log a_j + (N - a_j) \log (n - a_j)$ is symmetrical with respect to binary data in that the value for the expression is 2 log 2 for the two entities identical with respect to the scores zero or to unity. This is because in such comparisons N is 2, in which circumstances if a_j is 2 then $N - a_j$ contributes nothing to the expression and if a_j is 0 then it contributes nothing to the value of the expression and $N - a_j$ becomes 2.

It should be stressed that with the information measure designed to give equal weight to attributes scored as zero and unity, N refers to the number of entities (sites or taxa in normal analyses) and not the total number of individuals as with the information measure proposed for meristic data. Possibly one of the greatest advantages of information measures of dissimilarity is that they are readily adapted to handle mixed data. When this is done, the approach is similar to that using binary data; in other words, ΔI values are computed directly from data. Gain of information on fusion is computed separately for each of four types of data: binary, disordered multistate, ordered multistate, and continuous. The method of handling binary data has been outlined above. Continuous data are converted to eight grades and treated as ordered multistate. The principles of determining information gain in these cases are given in the Appendix.

Another advantage of information measures is in the effective handling of the problem of missing data. This is done by associating the information gain (ΔI) at each fusion cycle, including the first, with the number of degrees of freedom n, which are related to the number of attributes common to the entities being fused. Clifford *et. al.* (1969, p. 121) state: "The decision function is $2\sqrt{\Delta I} - \sqrt{(2n-1)}$, which is approximately distributed as zero mean and unit variance."

Groups are now fused on the basis of the minimum value of the decision function which scales the information gain ΔI in such a way as to delay fusions depending on few attribute comparisons. Hence the problem of poorly defined entities being involved in early fusions is avoided.

The manner in which this is achieved may readily be demonstrated by reference to the decision function. When entities are identical, ΔI is equal to zero and so the value of the decision function then depends entirely on n, the number of possible comparisons or more properly the number of degrees of freedom involved. When n is large the value of $-\sqrt{2n-1}$ will be algebraically less than when n is small and so fusions involving many comparisons will always take place before these involving few comparisons.

As a particular example consider the information gains and decision functions resulting from the fusion in pairs of the four entities listed, together with their attributes in Table 6.6. Here the information gains resulting from the fusions of A to B and C to D are the same, namely $\Delta I = 2 \times 2 \log 2$ which for logarithms to the base e is 2.77. Only two comparisons are possible for A and B since nothing is known as to the score of four of their attributes. In contrast to the similarity of information gains resulting from the fusion of the two pairs, the accompanying decision functions are quite different. Thus

$$\text{Decision function (A + B)} = 2\sqrt{2 \times 2 \log 2} - \sqrt{(2 \times 2) - 1} = 1.59$$

and

$$\text{Decision function (C + D)} = 2\sqrt{2 \times 2 \log 2} - \sqrt{(2 \times 6) - 1} = 0.01$$

Calculation of the other possible fusions will reveal that (C + D) has the

TABLE 6.6

FOUR ATTRIBUTES AND SIX ATTRIBUTES; BINARY DATA[a]

	Attribute					
Entry	1	2	3	4	5	6
A	0	*	*	*	*	0
B	1	*	*	*	*	1
C	1	0	0	1	1	0
D	0	0	0	1	1	1

[a] Missing attributes indicated by asterisks.

lowest decision function among these and so these two entities would fuse before all others. That is, the fusion of the poorly defined entities will have been delayed.

F. PROBABILISTIC MEASURES

A probabilistic similarity index that is able to incorporate all types of data has been proposed by Goodall (1966a). It depends on a knowledge of the frequencies of the attribute states among the entities and so is based on a knowledge of the total population to be classified. For each pair of individuals the exact probability is calculated so that a random sample of two will resemble one another not less closely than the two under test. Thus if an attribute has three states x_1, x_2, and x_3 with frequencies 0.1, 0.2, and 0.7, agreement between entities with respect to their both possessing the state x_1 would indicate greater similarity than their joint possession of state x_3. Having calculated the probabilities for each attribute separately, these are then combined assuming them to be independent.

The resulting combined probability P_t may be regarded as an index of dissimilarity ranging from zero for pairs of identical entities to approaching unity for completely dissimilar entities. The complement of P_t namely $1 - P_t$ is an index of similarity.

The χ^2 discussed earlier as a similarity measure (Chapter 6, Section A) may also be regarded as a probabilistic measure if its magnitude is associated with the degrees of freedom in a 2×2 or other 2-way table. This association will enable an estimation to be made of the observed value arising by chance had the attributes been independent. Likewise, if it is assumed the attributes are independent of one another it is possible to estimate the probability of an observed information gain (ΔI) arising by chance. In these circumstances $2\ \Delta I$ is approximately equal to χ^2 with the number of degrees of freedom involved equaling the number of attributes.

G. FURTHER PROPERTIES OF SIMILARITY MEASURES

In view of the large number of similarity measures proposed it is surprising there are so few in regular use. The remainder have been neglected largely because they possess special properties appropriate only to the problems for which they were derived or because they have undesirable algebraic properties over part of their range. From these observations it might be assumed that the commonly employed measures are sufficiently similar to one another to be used interchangeably, but as indicated earlier this is not so.

Consider, for example, two taxa identical in respect to all their proportions but differing in size. If the measure of similarity employed were Euclidean distance the two taxa would clearly differ, but if their similarity were measured in terms of the complement of the correlation coefficient, $(1 - r)$ interpreted as a distance, they would appear to be identical.

Hence, when choosing a similarity measure it is important that its features be known in order that they may be adjudged suitable to the problem in hand. While certain general characteristics have been given of the measures already described, these have mostly concerned scaling, inclusion or otherwise of double negative scores, nature of the data, etc. Here further properties are given against which the suitability or otherwise of the measure might be evaluated if it is being regarded in any way as a distance measure. These are a standard series of criteria to determine whether or not it might be regarded as a *metric*, this being a function that satisfies the four requirements (Kelley, 1955) discussed below. The properties of several measures as so tested are discussed in some detail by Williams and Dale (1965).

i. *Symmetry*. Given two entities (x, y), the distance (d) between them satisfies the requirement

$$d(x, y) = d(y, x) \geq 0$$

That is the distance between x and y is independent of the direction in which it is measured and must be positive provided the two taxa are not coincident.

ii. *Triangular Inequality*. Given three entities (x, y, z), the distances between them, $d(x, z)$, $d(x, y)$, $d(y, z)$, satisfy the requirement

$$d(x, z) \leq d(x, y) + d(x, z)$$

That is the length of any side of a triangle is equal to or less than the sum of the other two sides.

iii. *Distinguishability of Nonidenticals*. Given two entities (x, y):

$$\text{if } d(x, y) \neq 0, \text{ then } x \neq y$$

iv. *Indistinguishability of Identicals*. Given two identical elements (x, x) the distance (d) between them is zero, that is,

$$d(x, x) = 0$$

These properties are all clearly useful, and measures which fail to satisfy any of four criteria listed should be regarded with caution. As shown above, the correlation coefficient or rather its complement is not fully metric in that it fails to distinguish between nonidenticals differing in size but not shape That is, it fails to satisfy requirement (iii) and it furthermore may fail to meet item (ii) which concerns the triangular inequality. Likewise the coefficient proposed by Fager and McGowan (1963) is not fully metric since it fails to satisfy the symmetry requirement. Here the asymmetry is generated by the second part of the coefficient which varies according as to whether x is compared with y or vice versa. Other coefficients that are not fully metric in that they may fail to satisfy the triangular inequality are D^2 (Euclidean distance-squared) and the Jaccard and Czekanowski coefficients considered as their dissimilarity complements.

It does not follow that all three of these dissimilarity measures will necessarily fail to satisfy the triangular inequality requirement for a given set of data. Consider the data of Table 6.7 where three entities are specified in terms of ten binary attributes. From the "distances" between the entities as given in Table 6.8 it is clear that whereas D^2 and the complement of the Jaccard coefficient here satisfy triangular inequality, the complement of the Czekanowski coefficient does not meet this requirement.

Many of the other commonly used dissimilarity measures are metrics.

TABLE 6.7

BINARY DATA ON TEN ATTRIBUTES IN THREE ENTITIES[a]

	Entities		
Attributes	A	B	C
(10)	1	1	0
	1	1	0
	1	1	0
	1	1	0
	1	1	0
	0	0	1
	1	0	1
	0	1	0
	1	1	0
	1	1	0

[a] No missing data.

TABLE 6.8

DISTANCES BETWEEN ENTITIES OF TABLE 6.7 USING DIFFERENT DISSIMILARITY MEASURES

Comparison	Complement of Jaccard	Complement of Czekanowski	D^2	D
A − B	0.22	0.13	2	1.41
A − C	0.89	0.80	8	2.81
B − C	1.00	1.00	10	3.16

These include Euclidean distance, the simple matching coefficient, the Canberra metric, and information space. It is thought that the Bray-Curtis measure is not a metric (Lance and Williams, 1967b). However, for the majority of the less frequently used coefficients, it is not known whether or not they are metrics. If they are not, it does not preclude their use, but it is a worthwhile precaution to test their metric properties over the range of data which is employed.

Where there are missing data which often happens in taxonomy (see Chapter 5, Section B,3 for missing attributes), measures that are fully metric when data are complete may now become nonmetric especially with respect to the triangular inequality. Consider the binary example (Table 6.9), in which the asterisks indicate missing data and which we

TABLE 6.9

BINARY DATA ON TEN ATTRIBUTES IN THREE ENTITIES[a]

Attributes	Entities (taxa)		
(10)	A	B	C
	0	*	1
	0	1	*
	1	1	1
	0	0	0
	0	1	1
	1	0	*
	1	0	0
	1	1	1
	0	0	0
	0	0	0

[a] Missing data indicated by asterisks.

have assumed that granted full data, taxa B and C would have been identical.

Using Euclidean distance as a dissimilarity measure we obtain Table 6.10. From this table it is clear that the metric requirement of triangular inequality is met when the data are complete but not when the data are incomplete (AB > AC + BC).

Such a breakdown in the metric properties of Euclidean distance (D) resulting from a lack of complete data indicates care should be exercised in interpreting results using this or any other dissimilarity measure as a basis for either clustering or ordination. Taxonomists in particular should be alert to this loss of metric properties with incomplete data for of necessity they frequently encounter the occurrence of consequential attributes as well as missing attributes, for example, where states can not be determined due to the lack of suitable material.

Clustering procedures are most vulnerable to incorrect fusions in the early stages and in the example under discussion A and C appear to be more alike than A and B considering incomplete data though B and C are in fact identical. With ordination a breakdown of the metric properties of measures such as D may lead to the production of matrices with negative latent roots and so vastly complicate the interpretation of the results. However, in practice negative latent roots seldom appear in ordination and so it no longer appears essential as it once did for similarity measures to be metrics.

Several of the binary dissimilarity measures in common use consider only three of the four cells of a 2 × 2 table, and neglect conjoint absences—examples are the complements of the Jaccard and Czekanowski coefficient. The denominator of these coefficients varies according to the pair of entities under comparison and, hence, the spatial dimensions of the

TABLE 6.10

EUCLIDEAN DISTANCES BETWEEN ENTITIES OF TABLE 6.9

Comparison	Data complete[a]	Missing data
A − B	2.24	2.00
A − C	2.24	1.73
B − C	0.00	0.00

[a] Assumes no missing data and that B ≡ C.

unit hypercubes within which the results may be expressed are not the same over all comparisons. Similar considerations apply to the Bray-Curtis dissimilarity measure and to that version of the Canberra metric in which conjoint absences are eliminated from the scaling factor (see Chapter 6, Section B,2). In such cases, particularly if they involve ordination, a choice might be required between measures which are more mathematically acceptable in their preservation of spatial properties and those which produce the most meaningful results. Ecologists when confronted with a site/species matrix with a high proportion of zero entries are unlikely to accept as meaningful site similarities based on double negatives, however the alternative may distort the dimensions of the hyperspace.

7

Reduction, Transformation, and Standardization of Data

A. GENERAL

Ecological data in particular are often manipulated prior to the determination of measures of similarity and dissimilarity, and because the word "manipulated" has overtones of chicanery the reasons for this manipulation ought to be clearly indicated and their effects on the final classification ought to be understood.

There are many kinds of manipulations and, apart from data reduction, there is lack of agreement over terminology. To some recent authors it appears that these remaining operations are all "transformations" (e.g., Noy-Meir, 1973; Sneath and Sokal, 1973). We distinguish between transformations, standardizations, and scaling, each of which will now be discussed.

Transformations are alterations to the attribute scores of individuals without reference to the range of scores within the population as a whole. Thus without knowing the range exhibited by the entities in the population, it might be decided to replace all raw scores (n) by their square roots ($\sqrt{n}$) or logarithms ($\log n$) in order to satisfy some viewpoint on the nature of the data. This in effect is equivalent to making all the original measurements with an instrument graded in $\sqrt{n}$ or $\log n$ instead of n, and so no one score is influenced in magnitude by the others.

Standardizations are alterations which can not be made until an array of scores is available; these depend on the properties of the total array of values under consideration. An example is to divide an individual score by the summated score of the array (this is standardization by arithmetic total). We include under standardization such procedures as centering and normalization, which Noy-Meir (1973) regards as something different. His use of the term "standardization" is more restricted than ours.

The position is further complicated by the fact that standardization (in our sense) can occur at two stages. The first is the data prior to calculation of dissimilarity measures, and the second is during the calculation of these measures. Any measure which incorporates a fraction is standardized by virtue of possessing a denominator.

This applies to the Bray-Curtis dissimilarity measure, but in a somewhat peculiar fashion. The Bray-Curtis measure involves pairs of entities, and standardization is restricted to the pair under consideration and does not involve all entities. Hence, the basis of standardization varies from comparison to comparison. This may be confirmed by reference to the following tabulation concerning three sites each possessing three species, the numbers of each being recorded for each site.

	Site		
Species	A	B	C
1	10	6	8
2	3	2	9
3	20	16	15

The Bray-Curtis dissimilarity between sites A and B on the one hand and A and C on the other are as follows:

$$\text{A and B} = \frac{10 - 6 + 3 - 2 + 20 - 16}{10 + 6 + 3 + 2 + 20 + 16} = \frac{9}{57} = 0.16$$

$$\text{A and C} = \frac{10 - 8 + 9 - 3 + 20 - 15}{10 + 8 + 3 + 9 + 20 + 15} = \frac{13}{65} = 0.20$$

It will be observed that the denominator differs according to the comparison being undertaken.

The standardization employed in the Canberra metric dissimilarity measure is even more peculiar in that there is a separate basis of standardization not only for each pair of entities being considered, but also for each pair of attributes. Referring to the above data, the Canberra metric dissimilarity between sites A and B, and A and C are as follows:

$$\text{A and B} = \frac{1}{3}\left(\frac{10-6}{10+6} + \frac{3-2}{3+2} + \frac{20-16}{20+16}\right) = \frac{1}{3}\left(\frac{4}{16} + \frac{1}{5} + \frac{4}{36}\right) = 0.19$$

$$\text{A and C} = \frac{1}{3}\left(\frac{10-8}{10+8} + \frac{9-3}{3+9} + \frac{20-15}{20+15}\right) = \frac{1}{3}\left(\frac{2}{18} + \frac{6}{12} + \frac{5}{35}\right) = 0.25$$

It will be observed that the denominator differs with each pair of attributes under consideration. It should also be noted that the sum of fractions is divided by the number of fractions (in this example). This is an example of scaling (see below).

Scaling is a different but relevant procedure which we shall not elaborate on in this chapter. It is applied to measures of dissimilarity involving different numbers of attributes and reduces them to the same basis of comparison. Such scaling is particularly important with taxonomic data where the numbers of comparisons available often varies from species pair to species pair due to missing data. It can be applied to all unconstrained dissimilarity measures including Euclidean distance (D and D^2), the Shannon and Brillouin diversity measures, and is an integral part of the Canberra metric measure.

We now consider the various types of data manipulation primarily from an ecological viewpoint and follow with a brief consideration of data manipulation in taxonomy.

B. DATA REDUCTION

There are three reasons why data reduction may be desirable. The first is to reduce the number of computations and hence the resultant expense. The second is that it may permit the use of certain classificatory strategies which would not otherwise be available because of the mass of data. (Those inexperienced in the use of computer programs for classification have a misplaced confidence that computers can handle unlimited amounts of data. In ecology matrices of meristic data involving over 700 samples and 300 species can be obtained fairly easily and W. Stephenson has knowledge of several. Many classificatory strategies can not handle data of these dimensions even using the largest computers without the cost becoming prohibitive.) The third reason for data reduction is that if data show little or nothing of biological meaning there is no point in including them.

We now consider data reduction in an ecological context with most examples from marine benthos, and begin with the elimination of species. Removal of rarer species has been undertaken by a large number of workers, but there are several obvious circumstances when it should not be attempted. The first is when an interest has been declared in diversity. If diversity is measured in the simplest terms, for example, number of species per site, all species contribute equally irrespective of rarity on a per site basis. Even when Shannon or Brillouin diversities are involved, the rarer species, including "singletons," should be retained.

The reason is readily appreciated by reference to the formula for measuring the Shannon diversity. It will be recollected that diversity H is defined as:

$$H = N \log N - \sum_{1}^{S} n \log n$$

If a species with a single recording is rejected, the first term of the expression is reduced to $(N - 1) \log (N - 1)$ but the portion within the brackets in unaffected since 1 log 1 is always zero. The difference in magnitude of H through neglecting a single species with a single individual may or may not be great depending on the numbers of individuals of other species, but the difference in magnitude of the initial term of the expression is (2.4–6.0) over the range $N = 10$–100. To take an example, for $N = 100$, $N \log N = 461$, and for $N = 99$, $N \log N = 455$ (logs to base e). Clearly neglect of the single species here reduces the diversity by 6 units.

Notwithstanding the above, if species are to be eliminated, and this is usual practice, the first to consider are those species represented by single individuals. These contribute very little to the recognition of a site group; there is in fact an a priori likelihood that they could occur at any site. Frequently more than one of these once-only species occur at a given site, and as has been noted (Stephenson and Williams, 1971, p. 34) . . . "It is, in fact, often the case that a single station is unusually productive and contains an undue proportion of once-only species. It is seldom, however, that the ecologist is interested in the aberrant properties of individual stations; he is more commonly concerned with such similarities as do exist between aberrant and more 'average' stations. There is thus usually no objection in deleting all species which occur once only." The data on "singleton" species can sometimes give meaningful patterns by separate analysis, for example, by plotting occurrences on a map (see Stephenson *et al.*, 1970).

Typically very much more than once-only species are eliminated prior to ecological analyses. A variety of methods have been applied singly or in combination, and most have an arbitrary though quantitative basis. Several are almost purely arbitrary and depend on a ranking of the species by some measure which is easily calculated, followed by selection of a cut-off level. If data are binary the ranking can only be by number of sites occupied, i.e., by ubiquity. If data are meristic a measure based on number of individuals can be employed either in terms of the smallest recording at any given site or of total recordings at all sites. Examples of species eliminations of these kinds will be found in the following marine studies: Ebeling *et al.* (1970), Day *et al.* (1971), Field (1971), Hughes and Thomas (1971b), Stephenson *et al.* (1972), and Raphael and Stephenson (1972). In many cases even when meristic or continuous data were available, an ubiquity criterion was employed. Eliminations have ranged from a level of occurrence in ca. 2.5% of stations (Stephenson *et al.*, 1972) to a coverage of about 0.2 specimens per sample (Day *et al.*, 1971). Whatever basis of exclusion is used, it is clear that it ought to be of minimal severity and that some of the above workers may have exceeded this limit. The rationale of numerical analysis is that it is capable of handling considerable amounts of data, and if more data are excluded than necessary this rationale may be removed.

Several workers have employed an initial quite arbitrary exclusion level of low severity, followed by a more justifiable higher level exclusion, and a variety of methods have been employed. Most of these use a different intuitive or mathematical model from that used in the final analyses, partly because the final analyses can not economically be undertaken until after reduction has been effected. While this has been criticized by Williams and Stephenson (1973), it may be difficult to avoid; nevertheless it does give preference to classificatory strategies in which the same model can be used for reduction of data and then for classification (as in Williams and Stephenson, 1973).

A nonmathematical approach, but one with some intuitive appeal was used by Stephenson *et al.* (1970). In this case area patterns were sought from site recordings and as a preliminary to second stage data reduction the less ubiquitous species were map-plotted. This was followed by visual attempts to see any kind of single-species patterns in the data. Species occurring at less than 8 sites out of 400 were excluded because below this level no patterns were evident in a majority of cases.

Another method is to obtain measures of the strength of species/species linkages and to eliminate all species which fail to obtain a given threshold value in any comparison. Because in essence the exclusion of attributes from a "normal" analysis is now depending on the results of a preliminary analysis of an "inverse" nature, the general rationale is doubtful. It has more validity when applied to binary data than with meristic, because, as shown later the latter appear to require species standardization for species/species similarities or dissimilarities to become entirely meaningful. It was used with apparent success in the analyses of dredged catches of benthos by Stephenson *et al.* (1970). Two measures of species/species affinity were used for binary data, these were the product–moment correlation coefficient and the Czekanowski coefficient (by error named the Jaccard coefficient).

Yet another method involves analysis of a considerable mass of data by a cheap-running computer technique which can handle this mass as a preliminary to more expensive techniques requiring smaller amounts. A divisive monothetic analysis based on information theory and using binary data has been used by Stephenson *et al.* (1970, 1972). Species elimination depends on the degree of coincidence of distribution of the less ubiquitous with the more commonly recorded species. At each site group dichotomy, one group is delineated by the absence of a "key" species and a final site group is negatively characterized by an absence of a sequence of "key" species. Similarly, as a result of species groupings there is a final species group which is negatively characterized by a group of "non-key" sites. The species in this group are not linked in species-groups with the remainder and can be eliminated.

Methods which use one model for data reduction and another for classification, may well be open to criticism. One method of species reduction which uses the same model as that of the classificatory strategy was employed by Williams and Stephenson (1973) and in the amplified application by Stephenson *et al.* (1974). The classification involved variance partitioning and the contribution of each species to a designated entity/species matrix was computed. This still required an arbitrary cut-off level but in this case it was based upon contributions to the classification which was used.

The most recent refinement of species reduction techniques is to obtain an entity classification (e.g., site groupings) using one or more of the methods previously selected. Species which remain at "random" with respect to the entity groups can then be identified by a conformity test (see Chapter 9, Section B) and rejected in subsequent analysis. In addition, species previously rejected can be tested and incorporated in these later analyses. Trials have been made to date by Williams and Stephenson (1973) and Stephenson *et al.* (1974). Current investigations (involving W.S.) indicate that while arbitrary cut-off levels had been at approximately the current overall levels, they had excluded an important proportion of "worthwhile" species and so the basis of such "cut-off" levels requires further investigation.

Further species are sometimes eliminated because of the virtual impossibility of their being identified in a finite time. Even in areas where the benthic fauna is well known, it is usual to eliminate such groups as nematodes and ostracods (e.g., Field, 1971). The alternative to elimination is to "lump" and accept polyspecific taxa. Few would willingly agree to this because habitat or niche prescription of different species is an important goal of many ecologists. Bradbury and Goeden (1974) have taken the opposite viewpoint and in dealing with diversity of species in coral reef environments have shown that little is lost by dealing with taxa at approximately the family level.

We now briefly consider the elimination of entities from classification. This is usually performed on "obvious" grounds. A taxonomic example is the exclusion of a specimen so damaged that most of its characters cannot be recorded. An ecological equivalent is to discard (and repeat if possible) an attempted benthic sampling when the grab has failed to operate properly. Experience shows that such cases are not always as "obvious" as they seem. For example, when operating grabs on hard grounds containing shell fragments, a succession of catches may be discarded before one is finally selected. This may bias data by selection of maximal catches.

Sometimes inadequacy of the collecting technique is revealed following classification of data. For example, in one study with which we have been associated, certain species-impoverished samples segregated out in a preliminary analysis. It was then noticed that these had been obtained by use of a coarser sieve than for the remainder and so these samples were eliminated.

C. DATA TRANSFORMATION

In taxonomic classification by numerical methods, most of the raw data have usually a similar weighting—typically most are in binary form. Some of the most vociferous objections to numerical taxonomy have come from classic taxonomists who object to this uniformity. In ecological classification the problem is the opposite one. The raw data (species in sites) are not uniform because some species are more abundant and some less so. If raw data are used in some ecological analyses, abundance can be "overstressed" and the results can be dominated by a few extremely high values. For example, if an Euclidean distance measure of dissimilarity is used, the squaring of differences greatly accentuates the difference between abundant and rarer species.

Before analyses are begun, it is politic to make some sort of judgment on the importance to be given to abundance. Choices can then be made on the measurement of dissimilarity and transformation to apply (if any be needed); it is important to realize that dissimilarity measure is related to the transformation applied to the data.

A total judgment on the importance of abundance in the analysis cannot be made without the fuller consideration of interpretation of the results of analyses which are given in Chapter 9. At this stage we can merely say that two factors are involved; the nature of the data and the objectives or biases of the investigator. With respect to data, in some cases there are many codominant species of approximately equal abundance, or expressed alternatively, a high diversity. Here optimal ecological sense may be obtained by focusing attention on the abundant species at the expense of the rarer ones. In other cases the data may be "spikey" with occasional very high recordings, and here it may be desirable to reduce the importance of abundance. At the present time, assessments of the nature of the data are made by visual inspection and it is clearly desirable that this should be superseded by more objective methods.

The biases of the observer relate to the types of patterns which might be sought in the data, and this implies a choice between preferences. In some cases preference may be given to discrete groupings of attributes (species) in entities (sites). Choices of techniques may then be quantified by the number of attributes which attain a high and specified level of constancy and/or fidelity of attributes in entity groups. Stephenson *et al.* (1972) gave an example of this type of approach. In other cases, preference may be given to the technique which gives the best reflection of extrinsic (abiotic) attributes in the entity groups. It is then found that techniques which give stress to dominance at the partial expense of constancy and/or fidelity of biotic attributes are likely to be preferred (Raphael and Stephenson, 1972).

We next consider some commonly used dissimilarity measures in relation to transformation. As shown above (Section A), the Canberra metric measure is insensitive to large values and can be used with very "spikey" data either with or without a relatively mild transformation (e.g., $\sqrt{n}$). The outcome of the analyses will be biased toward presence/absence data and to the constancy and fidelity of species in site groups. In contrast, the Bray-Curtis measure is sensitive to larger values and its use is tantamount to a declaration of interest in dominance. If attribute values cover a very wide range, a tolerably drastic transformation (e.g., log $(n + 1)$ may be desirable. Euclidean distance types of measures, which include D^2 and variance measures, are particularly sensitive to "spikey" values, and transformation is almost mandatory to produce optimal ecological sense.

In past marine ecological work, four different transformations have largely been used. These have been $\sqrt{n}$, $\sqrt{n + c}$, $\sqrt[3]{n}$, and log $(n + 1)$. They should be related to their respective contexts as contained in such papers as Field and Macfarlane (1968), Field and Robb (1970), Ebeling *et al.* (1970), Day *et al.* (1971), Field (1971), Thorrington-Smith (1971), Stephenson and Williams (1971), Stephenson *et al.* (1972, 1974), Raphael and Stephenson (1972), and Williams and Stephenson (1973).

Attempts have been made to justify the choice of various of these transformations by reference to standard statistical procedures. In these, prior to use of tests of significance it is often necessary to transform the data until they are normally distributed. Extensive literature has arisen on this aspect of transformation, including Preston (1948), Patrick *et al.* (1954), MacArthur (1960, 1969), Patrick and Strawbridge (1963), Macfadyen (1963), Dahl *et al.* (1967), Cassie and Michael (1968), and Edden (1971).

It remains uncertain whether the transformation required to produce normality of data is also the transformation which will produce optimal ecological sense. A priori one would not expect that such a simple relationship would hold, if only for the reason that the stringency of the transformation desired is not a fixed thing but depends first on the dissimilarity measure to be employed and second on the various concepts of "optimum ecological sense." It seems from a tentative (and unpublished) study by Williams and Stephenson that optimal classificatory "sense" is generally obtained by using a weaker transformation than that required to transform data to normality.

The properties of the $\sqrt{n+c}$ and $\log(n+1)$ transformations require especial mention. The former was used by Ebeling *et al.* (1970), and the latter by Field and his co-workers, both in a marine context. With both transformations, the importance of high values is reduced, and that of small values is increased. The changed importance of the smaller values applies particularly when n is less than unity. With a $\log n$ transformation a negative value is obtained, and it is primarily to avoid this that $\log(n+1)$ is used.

Allied to the replacement of $\sqrt{n}$ by $\sqrt{n+c}$ is the replacement of n by $n + c$ in certain circumstances. This was done by Stephenson *et al.* (1972) when using the Canberra metric measure with marine data. With these data, pairs of attributes usually include many 1/0 recordings and also a few large numbers. Consider the data in the following tabulation.

	Sites	
Species	A	B
1	0	1
2	0	1000
3	100	1000

The Canberra metric intersite dissimilarity as estimated from these data is

$$\frac{1}{3}\left(\frac{1-0}{1+0}+\frac{1000-0}{1000+0}+\frac{1000-100}{1000+100}\right)=0.94$$

It offends common sense that species 1 contributes as much to the dissimilarity as species 2, and more than species 3; moreover, this remains true with most transformations. To avoid the Canberra metric measure being overbiased by 1/0 recordings, when there is a zero/non-zero comparison it seems desirable to replace the zero by a small number. Trial by the above authors showed that a value of 1/5 of the smallest entry in the matrix appeared satisfactory.

With the data given above, the intersite dissimilarity measure using this adjustment becomes

$$\frac{1}{3}\left(\frac{1.0-0.2}{1.0+0.2}+\frac{999.8}{1000.2}+\frac{900}{1100}\right)=0.83$$

In this example, the reduced value of the dissimilarity measure is not large; with more realistic data containing many 1/0 recordings, the difference can be considerable.

Taylor (1961, 1971) introduced flexibility into the range of transformations which may be applied to biological data by the use of a series of fractional powers of n, both positive and negative. Williams and Stephenson (1973) essayed the use of $n^{1/3}$ and $n^{1/4}$ in a classificatory context and after trial adopted the first. Realization that there is a wide range of options between and beyond $\sqrt{n}$ and $\log (n+1)$ increases the choice of transformations available and increases the problems associated with the selection of the correct choice.

Minor assistance might be gained by reference to the basic ways in which ecologists think about numbers, and particularly in the methods used for the grading of data. For example, Field (1971) has noted abundance grades seem to involve an intuitive process of logarithmic transformation. Thus an "abundant" species is in its thousands, a "very common" one in its hundreds, and so on. This gives general support to a logarithmic transformation.

Another somewhat arbitrary approach being investigated by W. Stephenson is the use of transformation of the type $n^{1/2}$, $n^{1/3}$, and so on. Here, using benthic data originally in meristic form, raw values gave a range of ca. 1000. The range after $n^{1/2}$ transformation was 32 and after $n^{1/3}$, 10. It was felt that division of the range into 10 major groups was appropriate and the cube-root transformation was employed.

D. DATA STANDARDIZATION

The present abbreviated account may be supplemented by reference to Noy-Meir (1970, 1973) who used the concept of ecological distance in an ordination approach to the classification of some Australian vegetation. We will illustrate some aspects of standardization by reference to the site species matrix (A) (Table 7.1) in which the entries are numbers of individuals and will take as our measure of dissimilarity squared Euclidean distance (D^2). We shall first consider site/site relationships. The D^2 between sites 1 and 2 is

$$(2000 - 1000)^2 + (20 - 10)^2 + (5 - 0)^2 + (2000 - 0)^2$$

This exemplifies how the D^2 measure is dominated by large values in that the value of D^2 is little affected by the inclusion or otherwise of species 2 and 3.

We have already indicated two ways in which this domination can be avoided. These are the use of a different dissimilarity measure and/or transformation of the original data. There is yet a third way. Instead of dealing with the numbers of species in each site we can operate on the proportionality of the total number of individuals at the site contributed by each species. Expressing these data as proportions (where the total is unity), matrix A to two decimal figures becomes converted to matrix B (see Table 7.2). This procedure of dealing with proportionalities within each site is *site standardization by totals* or, in this context, standardization by column totals.

A fuller discussion of the effects of this standardization requires comparison of all intersite dissimilarities, for example, by two trellis diagrams (Chapter 8, Section A). However, it will be noted that the domination in

TABLE 7.1

DATA MATRIX A, FOUR SPECIES IN FOUR SITES

	Sites			
Species	1	2	3	4
1	2000	1000	500	—
2	20	10	5	—
3	—	5	—	—
4	2000	—	—	10

TABLE 7.2

MATRIX B

Species	Sites 1	2	3	4
1	0.48	0.99	0.99	—
2	0.04	0.01	0.01	—
3	—	0.00	—	—
4	0.48	—		1.00

matrix A of the two large values of species recordings in site 1 has been reduced and this appears reasonable. There are however difficulties associated with attaining this reasonableness. The difference between sites 2 and 3 has largely disappeared; this may be less acceptable because we may have wished that the considerable differences in populations should count for something. A bigger difference concerns site 4 where great stress is now given to the small numbers of species 4. Possibly the greatest objection to site standardization, whether by totals or any of the other methods to be described, is the prominence given to the records from species-impoverished sites.

We shall now consider species/species relationships. Again referring to matrix A, the D^2 between species 1 and 2 is $(1980^2 + 990^2 + 495)^2$ while D^2 between species 1 and 4 is much less, viz., $(0^2 + 1000^2 + 500^2 + 10)^2$. This appears nonsensical; it seems obvious that the patterns of species 1 and 2 are similar and have been obscured by the fact that species 1 is more abundant than species 2.

TABLE 7.3

MATRIX C

Species	Sites 1	2	3	4
1	0.57	0.29	0.14	—
2	0.57	0.29	0.14	—
3	—	1.00	—	—
4	0.99	—	—	0.01

The obvious solution is to deal with the proportions of the recordings of a given species in each of the sites—this is *species standardization by totals*, in this context standardization by raw totals. The resultant matrix C is illustrated in Table 7.3. Again only an abbreviated comparison of the effects of the transformation on the D^2 measure is given. Species 1 and 2 are now identical, which makes good ecological sense. Considerable stress is now given to the rarest species (No. 3) which only occurs in site 2 and the emphasized importance of rare species may be a serious objection to species standardization.

We can briefly evaluate the effects of both standardizations in relation to Euclidean distance squared. Species standardizations appear to be required to produce sensible measures of interspecific ecological distances; this seems the only satisfactory way of equating the site distributions of species of different abundances. The advantages of site standardizations are less compelling, and it appears that by using appropriate transformations and dissimilarity measures this may be avoided. It should be noted that in cases where a Euclidean distance or scaled Euclidean distance (variance) dissimilarity measure can not be avoided, site standardization may also be needed.

In the above discussion we have use D^2 as a dissimilarity measure; let us now consider the consequences of using the Bray-Curtis and Canberra metric measures. Referring to matrix A and sites 1 and 2 the Bray-Curtis dissimilarity is

$$\frac{(2000 - 1000) + (20 - 10) + (5 - 0) + (2000 - 0)}{2000 + 1000 + 20 + 10 + 5 + 0 + 2000 + 0}$$

Two differences as compared with the D^2 measure are apparent, first, that differences in the numerator are not squared (hence, reducing the importance of large values) and, second, the existence of a denominator. The latter, which constrains values to a maximum of unity, is in fact a form of standardization. Previously, we standardized sites individually by their individual totals, and then determined dissimilarities. With the Bray-Curtis the dissimilarities are standardized by the summed totals of the species records in the pair of sites under consideration.

Again referring to matrix A and sites 1 and 2 the Canberra metric dissimilarity is

$$\frac{1}{4}\left(\frac{2000 - 1000}{2000 + 1000} + \frac{20 - 10}{20 + 10} + \frac{5 - 0}{5 + 0} + \frac{2000 - 0}{2000 + 0}\right)$$

Again standardization is involved, this time using the totals of each species in the two sites separately. We now consider the Bray-Curtis and Canberra metric measures with respect to interspecies differences, referring to matrix A and the species comparisons 1/2 and 1/4, respectively. The differences are summarized in the following tabulation.

	Species	
	1–2	1–4
Bray-Curtis	0.98	0.27
Canberra metric	0.98	0.75

Compared with the D^2 measure the differences are reduced (as they must be with an upper limit of unity) but the "illogicality" of the relatively greater difference between species 1 and 2 remains. In general, it appears that even with the built-in standardizations of the Bray-Curtis and Canberra metric measures further species standardization as originally outlined may be desirable.

A great variety of standardization procedures exist. These include standardization by total, either of columns (sites) or rows (species) as just outlined; standardization first by sites and then by species or vice versa. Double standardization (by maximum) was practiced by Bray and Curtis (1957) in their paper on ordination of vegetation in Wisconsin and from which sprang the index bearing their name. A further set of possibilities involves standardization by $\sqrt{\Sigma(\text{numbers of values})^2}$ which Noy-Meir (1973) distinguishes as normalization using the "norm." Again this can be by sites, by species, or both (successively or simultaneously); this is the method preferred by Noy-Meir (1970).

The above methods involve standardization by size measures, others involve measures of variation. These include range, variance, and standard deviation. In the case of variance and standard deviation, such standardization only becomes meaningful when a reference point is prescribed, and clearly this is the mean. When values are referred to the mean by prescription of zero mean, this is called centering. A typical method is to express values as positive and negative units of standard deviations about a zero mean; this Noy-Meir (1973) refers to as centering followed by normalization by standard deviation.

Noy-Meir (1970, 1971, 1973) reviews previous examples of standardization. They include site standardization prior to site analyses using presence/absence data by norm (Ochiai, 1957; Orloci, 1967b) and by standard deviation (Williams and Lambert, 1961, Webb *et al.*, 1967); and using quantitative data by total and by norm (Austin and Greig-

Smith, 1968). Species standardization prior to site analysis was carried out using quantitative data by Orloci (1966), Gower (1966), and by Greig-Smith *et al.* (1967). Species standardization prior to species analysis has been carried out on presence/absence data by norm (Dagnelie, 1960) and by standard deviation (Goodall, 1953; Williams and Lambert, 1959; Yarranton, 1967) and on quantitative data by standard deviation (Goodall, 1954; Dagnelie, 1960).

Two-step standardization has been used by Cattell (1952) in psychology and by Bray and Curtis (1957) and Austin and Greig-Smith (1968) in plant ecology. Simultaneous double-standardization has been used by Benzecri (1969) and by Noy-Meir (1970).

Finally it should be noted standardization is built in to many of the similarity/dissimilarity measures not specifically discussed above. Thus the Jaccard and Czekanowski coefficients are standardized by the arithmetic mean, Kulczynski's second coefficient employs the harmonic mean and Ochiai's and the related Fager and McGowan indices standardize with the geometric mean.

E. REDUCTION, TRANSFORMATION, AND STANDARDIZATION OF TAXONOMIC DATA

The need to eliminate data in taxonomic studies arises less often than in ecological studies. Nonetheless it is often necessary to neglect diseased, damaged, or immature specimens.

The transformation and standardization of taxonomic data encounters most of the problems faced in ecology and the general approaches are similar. Size in individuals may be a reflection either of the genotype, or the environment, or both. Thus dwarf plants may result either from a given genotype or an unfavorable environment.

Various attempts have been made to eliminate the influence of the environment and often the most popular are the ratios of pairs of measurements. However, ratios should be used with caution unless it can be shown that the "line of best-fit" through the measurements passes through the origin.

The most usual transformation is to logarithmic form for it often happens that such a transformation allows for growth being exponential. Nonetheless other transformations have been employed just as in ecology. In their study of hybridization on *Eucalyptus*, Clifford and Binet (1954) transformed their data as follows: for fruit weight the cube root was taken, fruit length was expressed in logarithms, and the length of the peduncle was converted to square roots.

F. DISCUSSION OF DATA MANIPULATION

It will be evident that each variant of each type of data manipulation (reduction, transformation, and standardization) will give a different data matrix and hence is likely to give a different classificatory end point.

It should be equally evident that the types of data manipulation which should be invoked depend on prior ecological or taxonomic decisions. The consensus of ecological opinion in favor of data reduction suggests that the distinctly uncommon species can and should be neglected in ecological classification. Equally the widespread use of $\sqrt{n}$ and log n types of transformation and of standardizations suggests that high values of specific recordings should not be taken at their face value. This is particularly so in methods involving the concept of Euclidean distances, and hence has special significance to those who approach numerical ecology by ordination techniques.

Species standardization results in species which appear intuitively to be related in an ecological sense appearing to resemble each other more closely and, hence, if the ecological objective is to group together species in the best manner, species standardization is indicated. In passing, it should be noted that the usual objective appears to be grouping of sites and that this has become the "normal" classification. There is no certainty that this emphasis will continue, and stress on interaction of organisms springing from supraorganismal concepts of communities may well cause reorientation of objectives. To the dedicated biologist there may be a feeling of lése-majesty in using the all important organisms merely to delineate areas on maps or charts.

Species standardization prior to site classification has the effect of increasing the importance of relatively uncommon species and, hence, the classificatory importance of fidelity against constancy and dominance. In most situations, it appears that this is an undesirable effect, but could well become important in cases where the uniformity of ubiquitous and dominant species fails to give meaningful ecological subdivisions—these are the situations from which the Braun-Blanquet originated. There is an anomaly between data reduction by elimination of species and species standardization; below a given level species are discarded and above it they become particularly important.

Site standardization seems a useful preparation for site classification, but increases the importance of those species which are present in depauperized sites. Only by ecological judgment of the actual situation can the propriety of site standardization be decided. Such standardization has an intuitive appeal to many ecologists, particularly when there are doubts about the performance of gear in absolute terms.

8

Similarity Matrices and Their Analysis

A. VISUAL MATRICES—TRELLIS DIAGRAMS

A matrix consists (in the present context) of a series of measures of similarity or dissimilarity between the pairs of entities to be classified.

Matrices can be displayed in a numerical visual form. Taking a simple example, suppose we have four entities (I, II, III, and IV) and the following six values of interentity measures: I–II 0.5, I–III 0.3, I–IV 0.1, II–III 0.5, II–IV 0.3, and III–IV 0.7. These can be expressed as a two-dimensional matrix or "trellis-diagram" as follows:

	I	II	III	IV
I		0.5	0.3	0.1
II			0.5	0.3
III				0.7
IV				

Only the upper triangle is given, as this is clearly the mirror image of the lower triangle.

The simplest method of matrix analysis is visual. The continuous values of the measures are graded or ranked into a small number of arbitrary categories, and (assuming similarity measures are involved) in the words of Macfadyen (1963, pp. 194–196): "The indices are arranged in the squares of the trellis diagram . . . , the linear order of the samples being the same in the rows and the columns. This linear order is then shuffled round by trial and error so as to bring the highest numbers to the centre diagonal and the lowest numbers farther from it. A moment's thought will show that this results in samples with high affinity, as measured by the index, coming together. When arrangement has been carried out as far as possible it is usually found that the samples fall into discrete groups." We would disagree with this statement only in the words "usually found." The requirement for classifications which optimize discontinuities in the data suggests that these are frequently not easy to locate by visual analysis.

Now that clustering techniques using computers are available, the visual sorting should be regarded as obsolete or preliminary. Nevertheless visual analyses continue to be published and examples in marine ecology include Boudouresque (1970), Lie and Kelley (1970), Pearson (1971), Popham and Ellis (1971), Boesch (1971), Tenore (1972), and Gage (1972). Thorrington-Smith (1971) in marine phytoplankton work gave full matrices without grading or rearrangement.

Matrices in trellis form, by displaying the total relevant data in a simple way, assist in their familiarization and may be a guide to intuitive thinking. They served this purpose as a preliminary to data reduction in work by Stephenson *et al.* (1970).

B. CLASSIFICATORY STRATEGIES IN GENERAL

These strategies have already been considered in general terms. The alternatives which are available for the hierarchical methods are either *divisive* or *agglomerative* and also either *monothetic* or *polythetic.* These concepts have already been briefly introduced and are now expanded as being the bases of the strategies outlined in later sections.

Three possible methods exist, at least in theory; these are monothetic divisive, polythetic agglomerative, and polythetic divisive. Of these three, as pointed out by Williams (1971), the polythetic divisive is the most interesting. The programs have the economy of computing characteristic of other divisive methods (and, hence, can handle large data matrices), and because they employ the total data, avoid some of the misclassifications of monothetic methods. Unfortunately most of the current programs test only one possibility at each step, for to test all would reintroduce the need for extensive computing.

In practice there are few polythetic divisive programs readily available and so this procedure will be considered only briefly. Of the two common approaches one subdivides the initial population on the basis of a single attribute and then reallocates apparently misclassified entities on the basis of a maximum likelihood procedure; the other undertakes a principal component analysis (see Chapter 13, Section B) on the total population and then subdivides on the basis of the principal component scores on successive axes.

C. MONOTHETIC DIVISIVE HIERARCHICAL CLUSTERING METHODS

1. General

While the balance of advantages lies with agglomerative methods, monothetic divisive methods have a number of advantages as listed below.

1. Group definitions are simple and nonambiguous. Typically there are a series of yes/no alternatives; and the groups are defined in these strongly contrasting terms. The factor which is stressed is constancy and in all "yes" answers the constancy of the attribute is always 100% within the group.
2. The majority of groups remain stable as additional entities are added to the matrix. Alterations will occur, however, if a sufficient number of new entities affect the priorities in the choice of attributes.
3. Computation is relatively fast. This is because usually there is more interest in the upper than the lower levels of the hierarchy. In a divisive method the process can be halted at the required level, whereas in an agglomerative one, the lower fusions (which are often neglected) must be undertaken before the upper levels are attained. As Williams (1971) noted for a population of n individuals, agglomerative methods require the calculation of at least $(n - 1)^2$ fusions, which may involve both excessive computer time and expense.
4. At least in theory, divisive strategies begin classification when the total information available is a maximum. In contrast, agglomerative strategies begin with the information pertinent to only the pairs of entities (taxa or sites) and so depend on the accuracy of the inter-individual measures, where the possibilities of error are greatest.

Divisive methods are particularly suitable for handling very large data matrices. They can readily be employed in data elimination to reduce large matrices to a size where agglomerative methods can be employed. This we considered earlier (see Chapter 7, Section B).

Monothetic methods are liable to generate "misclassifications" for as Williams (1971) has stated:

> "We suppose two groups, X and Y, to have been separated monothetically on a binary attribute, possessed by X and lacked by Y. Consider then an individual A, which on the whole resembles the members of Y but happens to possess . . . the division-attribute. A will appear in X, and since most users are concerned with overall (i.e. polythetic) resemblance, will appear to the user to have been misclassified."

It should be borne in mind that if a monothetically based classification is required then the methods do not generate misclassifications, for they arise only when we expect monothetic divisions to produce groups with overall resemblance.

They hinge on careful choice of the first and successive attributes on which the entities are divided, and, hence, depend on the properties of the entity array. This careful choice of attributes has a familiar parallel in a parlor game in which after an original trichotomy ("Animal, vegetable, or mineral?"), there are successive dichotomies ("Can you eat it?") chosen to subdivide the array of possible objects as economically as possible. In the present types of classifications we look for attributes which divide the total array of entities into the two most dissimilar groups, and those which have the greatest internal homogeneity. Hence the chosen attributes are "keys" to the remaining attributes. Two methods have been employed for the determination of the division attributes—those depending on information theory and upon χ^2, respectively. The χ^2 measures were used first and have been superseded by the information theory measures, which have the advantage of being capable of handling mixed data (Lance and Williams, 1968a,b).

2. Strategies Based on Information Measures

The choice of the attributes on which to divide appears to be most effectively achieved by information theory measures. The total information content of the population is computed (I_C), and then the population is divided into two subsets using each attribute in turn to give I_A and I_B. The attribute selected as the basis of subdivision gives a maximal value for $I_C - (I_A + I_B)$; where this is done the two subsets are each as homogeneous as possible (high homogeneity means low heterogeneity, which means low information content). The process is repeated on each of the two subsets resulting from the first dichotomy, using the remaining attributes. Thus the attributes to be scanned declines by one at each division, but the number of entity groups involved doubles.

This procedure has been used successfully for biogeographical and ecological studies on birds (Kikkawa, 1968; Kikkawa and Pearse, 1969) and in benthic studies (Stephenson *et al.*, 1970, 1972) using the program DIVINF (Lance and Williams, 1971). The program operates at considerable speed but is restricted to binary data in which there are no missing values. The computing time with DIVINF depends on the square of the number of attributes multiplied by the number of entities, hence, it is particularly useful for ecological problems with many sites and few species.

A second divisive program operating on the same principles as DIVINF but capable of using disordered multistate, ordered multistate, and continuous as well as binary data is also available. Known as MULTDIV (Lance and Williams, 1971), it is the divisive analog of the agglomerative clustering strategy MULTBET referred to below (Section D,2). This program is slower running than DIVINF and is not suited to situations where the bulk of the data is quantitative.

Divisive programs in general are not favored in taxonomic studies because, as has already been stated, subdivisions of populations on the basis of single attributes frequently leads to the "misclassification" of some entities. It is to avoid this situation that reallocation programs such as REMUL * are being developed (see Section B) and have been applied with considerable success to the classification of a set of grass genera.

3. Strategies Based on χ^2

Prior to the application of information theory statistics to classificatory problems divisive strategies were largely based on χ^2 measures. These were designed initially for vegetation mapping (Goodall, 1953) and their application and properties have been investigated in particular by Williams and Lambert (1959, 1960, 1961) who described the technique as "association analysis." It is by this name that the technique is widely known. Their general method of operation is as follows.

Given a set of n binary attributes for a set of entities the χ^2 are calculated for all attributes taken in pairs. These are then summed over all attributes and that with the largest $\sum_1^n \chi^2$ is used as the basis for dividing the set of entities into two subsets—those possessing the attribute and those lacking it. With a two-state attribute one subset possesses the attribute in one state and the other subset possesses it in the alternate state.

When there are missing data the individual χ^2 are scaled by division by the number of entities before the summing of the resultant values to seek the largest. As described above the division of the population in two subsets is on the basis of the attribute associated with the largest value of χ^2 instead of information content. Likewise each subset is further subdivided in the same manner as the original array of data until the required number of subsets is obtained.

Programs for divisive χ^2 strategies based on binary data only have long been available and Williams and Lambert (1960) produced such a program for vegetation analyses. More recently, programs capable of handling mixed data and allowing for missing entries have become available (MULASS, Lance and Williams, 1968b). Though used widely in ecology,

* C.S.I.R.O., Div. Comput. Res., Canberra, Australia.

association analysis has been little used in taxonomy. Nevertheless MULASS was employed with advantage by Young and Watson (1970) to classify 543 genera of dicotyledons.

D. AGGLOMERATIVE POLYTHETIC HIERARCHICAL CLUSTERING METHODS

1. General

For reasons already stated, these are the strategies which have been most used in the recent past. They are of two types: those which are based on successive information gain or increase in diversity as groups are fused and those which are not. The latter may include many different measures of similarity/dissimilarity and many different clustering strategies.

2. Based on Successive Information Gain

These strategies are mostly used when the bulk of the data are binary and disordered multistate and in which there may be missing values. Hence, they have found considerable application in taxonomic problems. Clustering is based on minimal information gain (ΔI) at each fusion or on a minimum value of the decision function $2\sqrt{\Delta I} - \sqrt{2n+1}$ at each fusion cycle (see Chapter 6, Section E). Where data are binary and complete, the program CENTCLAS (Williams *et al.*, 1966) is available. It has been applied with considerable success to the classification of conifers by Young and Watson (1969).

When attribute data are mixed and there may be missing values, the information gain on fusion may be calculated as shown in the Appendix. The algorithms set out there are similar to those used in the programs MULTBET (Lance and Williams, 1967b) and INFO (Orloci, 1969). These methods have been applied successfully to the classification of grasses (Clifford *et al.*, 1969) and legumes (Edye *et al.*, 1970; Burt *et al.*, 1971). It was, however, rejected by Lavarack (1972) in his study of orchids and Correll (1974) in his study of sedges. Besides these taxonomic applications it has also been employed successfully by Conner and Clifford (1972) in a vegetation survey on North Stradbroke Island, off the Queensland coast.

Several properties of the two previous strategies should be carefully considered before they are adopted for any classificatory exercise. The first is that, because computing time depends on the square of the number of entities multiplied by the number of attributes, they are particularly suitable for studies involving few entities and many attributes. Second, the strategies are intensely clustering and produce "tidy-looking" dendrograms, which is possibly why they have been favored by some taxonomists. However, there are disadvantages associated with intense clustering and these we shall consider in Section 3,a below. The main disadvantages are group size dependence and a tendency to form groups of diverse members whose main property in common is their dissimilarity with other groups. Information gain programs are indicated when it is evident that data are weakly structured, and it is prudent that the relationship of the resultant clusters should be examined by ordination of the interentity dissimilarities as determined by another measure.

Information gain methods have been little used by ecologists. Without the problems of multistate data and missing values, there are few advantages in these methods. Moreover, the existing programs incorporate "double negatives"; that is acknowledge conjoint absences, and so require additional sophistication if these are to be avoided. The programs are at their strongest in dealing with binary data (which most ecologists would avoid) and at their weakest in dealing with meristic or continuous data (which are the data of greatest ecological value). A final disadvantage is that the information gain strategies which have been used refer to the original data at each stage in the fusion cycle. The remaining strategies do not refer to these data beyond the stage of determining the interentity dissimilarity measures and many reduce the volume of data to be considered as the process of clustering proceeds (see combinatorial strategies below).

3. Strategies Not Based on Successive Information Gain

a. General

These strategies begin their clustering from a matrix of interentity measures of dissimilarity and can be used in conjunction with any of the indices of dissimilarity. These include some which do not appear to have been used, for example, information gain between pairs of entities (see Chapter 6, Section E) or the dissimilarity measure $(1 - r)/2$ which can be derived from the correlation coefficient (see Chapter 6, Section C).

Because the majority of strategies have a readily conceived geometrical interpretations there are advantages if the measures used are metrics; however, this is not essential.

There are eight main clustering strategies available, some of which have been known for several years and have acquired a series of alternative names. The strategies are nearest-neighbor, furthest-neighbor, centroid, median, group average, incremental sum of squares, minimal variance, and flexible sorting. For a general account of these the reader is referred to Lance and Williams (1966b, 1967b) and Burr (1968, 1970).

Decisions on which strategy is to be employed may depend on properties of the strategies which only became apparent after some general experience had been gained in their use. In each strategy, different relationships develop between entities and clusters, or between one cluster and another as the analyses proceed. Though these relationships are often obvious where the basic algebra is examined in hindsight, they are often far from obvious until the strategy has been employed on several kinds of data.

Some strategies are "space-conserving," others "space-dilating," and the remainder "space-contracting" (Lance and Williams, 1967a).

If the original interentity measures are regarded as occurring in a given space, then sometimes the properties of this space remain unaltered as clusters form, but in other strategies the clusters alter the properties of the space near to them. Thus, "in a space-contracting system a group will appear, on formation, to move nearer to some or all the remaining elements; the chance that an individual element will add to a pre-existing group rather than act as the nucleus of a new group is increased; and the system is said to 'chain' [for a measure of chaining, see Williams, Lambert, and Lance, 1966]. In a space dilating system groups appear to recede on formation and growth; individual elements not yet in groups are now more likely to produce 'non-conformist' groups of peripheral elements" (Lance and Williams, 1967a, p. 374).

These concepts can be considered more pragmatically in terms of their clustering properties. Some strategies are weakly clustering, give chains of entities, and are not of any great conceptual value (these are "space-contracting") and are exemplified by the nearest-neighbor method. Other strategies, for example, incremental sum of squares (and the successive information gain strategies considered earlier), are strongly clustering and are of considerable conceptual value (they are "space-dilating"). Intermediate to these are such strategies as group average ("space-conserving"). The flexible strategy is unique in that it can be altered from space-contracting to space-dilating; in its usual operation it has become space-dilating.

The above paragraph implies that all the advantages are with strongly clustering methods, but this neglects the phenomenon of group size dependence which occurs when these methods are used. As a group becomes larger it becomes more difficult to join (it "recedes in space") and there is a greater tendency for a second group to form. This means that groups owe something to the interrelationships of entities, and something to the number of entities present. It is not possible in most cases, by drawing a horizontal line across a dendrogram to obtain groups with comparable characteristics, for example, of constancy of attributes. In particular, entities that have weak relationships to other groups and that would "chain" with them in space-conserving strategies are now segregated as a "non-conformist" group whose members are quite obviously dissimilar. This last may or may not be a disadvantage. There are some advantages in the approximate equality of group sizes, which facilitates tests of conformity of attributes (see Chapter 9, Section B). There is a parallel in human affairs with intensely clustering strategies. When a country club is forming it is easy to join, but when it becomes large, qualifications become more stringent, and there is a greater tendency for a new club to form in the area.

Another property of certain strategies is "nonmonotonicity" or a tendency to cause reversals in the dendrograms, and this is illustrated below under centroid clustering. Strategies such as this are to be avoided.

Finally, some clustering strategies depend on the total properties of groups prior to fusion (whether or not these groups are each composed of single entities). Others neglect group properties but effect fusion dependent on the properties of a pair of selected entities, one in one group and one in the other. There are obvious conceptual objections to these strategies which depend solely on interentity measures. There are also practical objections because throughout these clusterings (as in the case of successive information gain) all the original data are required. In the other cases, called by Lance and Williams (1967a) "combinatorial" to distinguish from "non-combinatorial" interentity single-linkage strategies, the original data can be discarded immediately a cluster is formed. This facilitates computing, reduces expense, and allows larger matrices to be accommodated. The eight clustering strategies referred to above will now be considered in turn.

b. Nearest-Neighbor

In all strategies fusions begin with the most similar pair of entities, as established by whatever dissimilarity measure is employed—to this extent all strategies begin with nearest-neighbors, and the differences appear in subsequent fusions.

In nearest-neighbor if a third site is nearer to *either* of the first two fused than to a fourth, it fuses with the first two. This can be illustrated by considering four entities with two attributes as in Fig. 8.1. Expressed as a dendogram the fusions would appear as in Fig. 8.2.

Nearest-neighbor is the oldest of the conventional strategies and one of the two "single-linkage" noncombinatorial methods. In a recent comparison of strategies Pritchard and Anderson (1971) used the words "least useful" of this (and another) strategy, and their dendrogram showed excessive chaining. It continues to be used, probably deriving from Sokal and Sneath (1963). In studies of animal biogeography it has been used by Holloway and Jardine (1968) and by Peters (1971). In marine ecology, Field and Macfarlane (1968) used it apparently satisfactorily in studies of the South African intertidal biota, and later Day *et al.* (1971) tried it (but preferred group average) in studies of the grab benthos off the North Carolina coast. Field (1971) also used this strategy, but not as the main one, in his benthic studies in South Africa. A recent example of its use by Thorrington-Smith (1971) in a study of phytoplankton gives as many as 12 species chained together. An earlier example by McConnaughey (1964) in plankton work appears to have been generally overlooked, obscuring the potential value of his coefficient of association. Nearest-neighbor clustering has also been used for the classification of *Rhizobium* ('t Mannetje, 1967a), *Trifolium* ('t Mannetje, 1967b), and grasses (Clayton, 1970).

From a utilitarian aspect, this strategy should be regarded as obsolete, but it has received unexpected support on mathematical grounds from

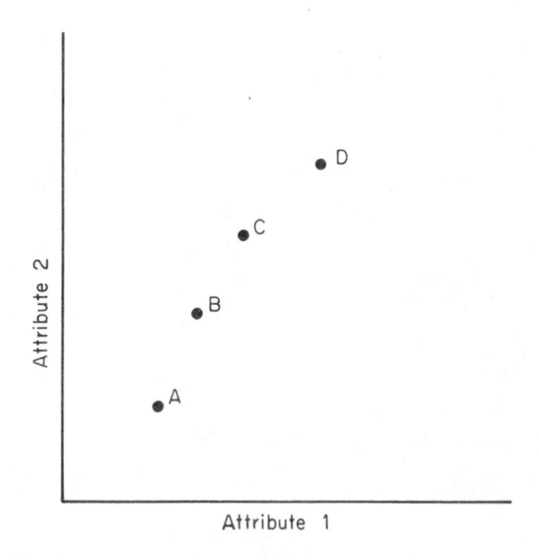

Fig. 8.1. The relationships between four entities A, B, C, and D with respect to two attributes.

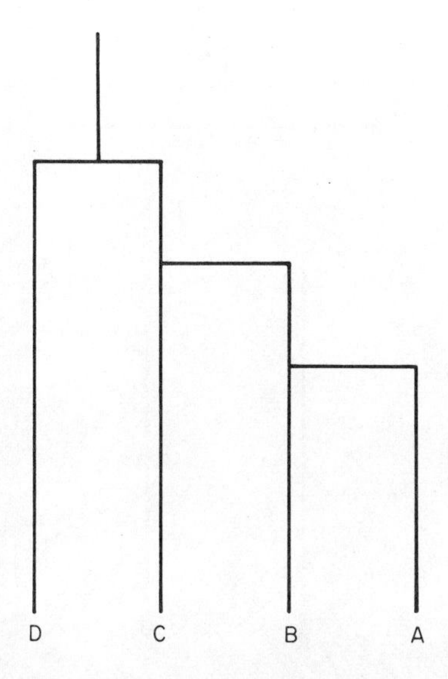

Fig. 8.2. The relationships between the same four elements A, B, C, and D expressed in terms of a dendrogram based on nearest-neighbor clustering.

Jardine and Sibson (1968). The criteria they believe should apply to classificatory strategies virtually confine one to nearest-neighbor fusion. A controversy arose between a "Cambridge school" and an "Australian school" over this and related issues, and this can be followed from Williams *et al.* (1971b), Sibson (1971), and Jardine and Sibson (1971a). It appears from Jardine and Sibson's (1971b) book that this controversy is abating and they say (Introduction, p. 10): "For many such problems it is known that there exists in general no unique optimal solution, and no computationally feasible algorithm is available which is guaranteed to find an optimal solution. The criterion by which the validity of application of a method of automatic classification to such problems must be judged is, simply, 'How well does it work?' "

c. Furthest-Neighbor

Like nearest-neighbor this is a noncombinatorial single-linkage method operating at an entity/entity level. It operates precisely oppositely to nearest-neighbor (as its name suggests) and fusions are based on the minimal distance between an entity and the most remote one in a group or between the most remote entities in two groups. Lance and Williams (1967a, p. 374) state: "Since on growth a group will recede from some

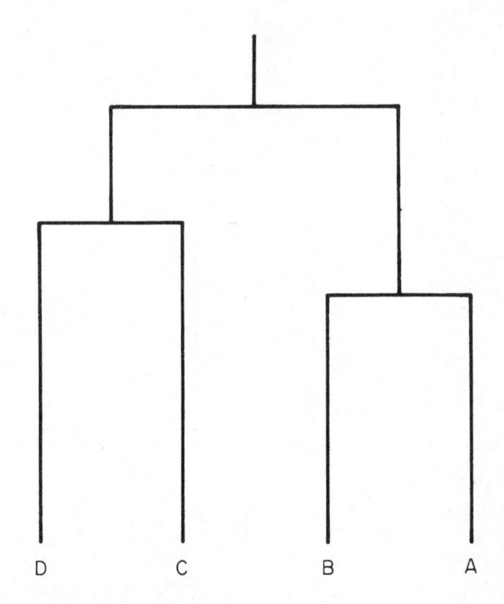

Fig. 8.3. The relationships between the same four elements A, B, C, and D expressed in terms of a dendrogram based on furthest-neighbor clustering.

elements and nearer to none, it is markedly space-dilating." In other words, it clusters intensely as is illustrated in Fig. 8.3 in which the data used in the previous example are shown as a dendrogram using furthest-neighbor clustering.

This strategy was suggested by Macnaughton-Smith (1965), but appears to have had few applications. It has been used for the classification of *Rhizobium* ('t Mannetje, 1967a) and *Trifolium* ('t Mannetje, 1967b). Following Jardine and Sibson's (1968) critique of classificatory methods and their obvious preference for single-linkage strategies it was tried recently by Stephenson *et al.* (1972) to analyze benthic grab data. Other methods were shown to be preferable for the problem in hand. Meanwhile, Pritchard and Anderson (1971) regard this as one of the more useful of the strategies used in their ecological studies of Scottish vegetation.

d. Centroid

Here, fusion of an entity into a group or fusion of pairs of groups depends on the coordinates of the centroids when considered in multidimensional space. The groups are fused on minimal distance between these centroids.

Lance and Williams (1967a, p. 375) have commented: "The centroid strategy is compatible for all coefficients and is space conserving. The ensuing simplicity of the overall model has not unnaturally tended to endear the system to users, but . . . the monotonicity requirement of Eqn. 2 is not met, and reversals, particularly with some measures, can be extremely troublesom (Williams, Lambert and Lance, 1966)." The existence of reversals is one of the problems which have appeared unexpectedly with the development of numerical methods. Lance and Williams (1967b) noted numerous cases using centroid sorting when after a dendrogram fusion the next fusion takes place at a lower level than the original, i.e., at lesser dissimilarity. An example is given in Fig. 8.4. Because dendrograms, however derived, are often interpreted divisively it appears that after a group divides into two, these are more dissimilar than the group generating them, which is quite absurd. In an agglomerative context they can be appreciated by reference to Fig. 8.5. Imagine three entities A, B, and C almost equidistant at two units apart, but with BC fractionally less than 2. B and C will fuse first to give point D. AD is now shorter than either AB or AC. Hence, the group after fusion is closer to the next entity it fuses with than the original "fusees" were with each other. More generally in an isosceles triangle with AB = AC, and with BAC = 2θ, it can be shown that inversions (reversals) will occur if $\cot^2 \theta$ lies between 3 and 4, that is, when the angle θ is between 30° and approximately 26°34′.

Because of the seeming illogic caused by reversals, it is clearly desirable to avoid methods which are nonmonotonic. For this reason centroid sorting has been largely avoided and the strategy may be regarded as obsolete.

Centroid sorting is described by Sokal and Sneath (1963) as the weighed pair and weighed group methods. It has been used by Hagmeier and Stults

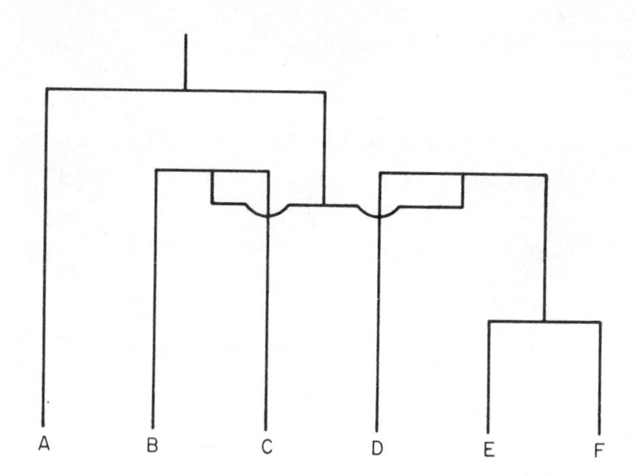

Fig. 8.4. A dendrogram illustrating the manner in which centroid sorting has resulted in an inversion following the fusion of groups BC and DEF (adapted from Williams *et al.*, 1966).

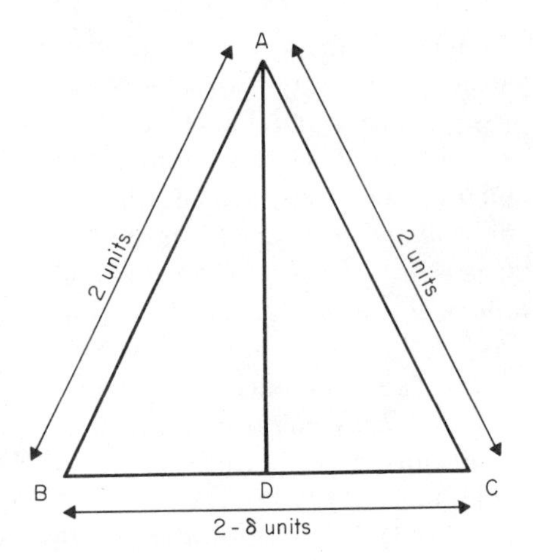

Fig. 8.5. An illustration of the manner in which centroid sorting may lead to inversions when three almost equally dissimilar entities are clustered. A, B, and C are the entities and D the product of the first fusion. The distance AD is now less than the length of any triangle side.

(1964) and by Hagmeier (1966) in biogeographical studies of North American mammals, by Ducker *et al.* (1965) for classifying green algae and by 't Mannetje (1967b) for classifying *Trifolium*, and El-Gazzar *et al.* (1968) for classifying *Salvia*. A recent application in ecology is by Popham and Ellis (1971) working on marine benthos.

e. MEDIAN

In centroid sorting if a small group fuses with a large one, it loses its identity and the new centroid may come to lie entirely within the confines of the larger group. To indicate the individuality of the smaller group it is desirable that the group obtained after fusion should be intermediate in position. This is effected in median sorting by regarding the groups as of unit size and obtaining an unweighted median position after fusion.

This strategy was apparently first suggested by Gower (1966) with a view to preventing large groups from dominating classifications to the exclusion of smaller groups. There are few reports of the use of the strategy possibly because it is prone to produce inversions as does centroid clustering. It has, however, been employed by Clayton (1972) for the study of grasses and Sands (1972) for the study of termites.

f. Group Average

This strategy can be illustrated through the entity/cluster situation, and here the mean of the distances of the entity to each constituent of the cluster is derived. Fusion is with the cluster giving the shortest mean distance. Similar considerations apply in cluster/cluster situations.

Lance and Williams (1967a, p. 375) state: "The system is less rigorously space-conserving than is centroid but, since it has no marked tendencies to contraction or dilation, it may be regarded as a conserving strategy." Hence, it gives only moderately sharp clustering, however, it has the advantages of being monotonic, little prone to misclassification, and with little group size dependence.

The method was originated in its entity/cluster form by Sokal and Mitchener (1958) who termed it the "unweighted group mean method"; it has also been called the "unweighted pair-group method."

Group average sorting is a generally satisfactory technique and has been used in ecological work in America (Jones, 1969; Ebeling *et al.*, 1970; Day *et al.*, 1971), South Africa (Field, 1970, 1971), Scotland (Pritchard and Anderson, 1971), Australia (Stephenson *et al.*, 1972; Raphael and Stephenson, 1972), and the Red Sea (Loya, 1972). In the Australian work (Stephenson *et al.*, 1972), it has been used in comparison with a more intense strategy, and there were few indications which produced the "better" result. Nonetheless being a weakly clustering strategy group average has an important role to play in that it can be usefully employed to check for misclassifications resulting from the application of more intensely clustering programs. In this regard its use was favored by Lavarack (1972) in his classificatory studies of orchids.

Later we shall refer to the unfortunate regionalization of availability of computer classificatory techniques. Because group average is generally satisfactory and is more widely accessible than many others, its use may well continue and extend.

g. Incremental Sum of Squares

Squares of Euclidean distances (D^2) are used as distance measures and after uniting the pair of elements whose D^2 is a minimum, subsequent entities are fused such that the sum of D^2 within a cluster increases by the smallest amount. Because the total sum of squares is constant, if the sum of D^2 within a cluster increases minimally, then it follows that D^2 between clusters is increased maximally. As both Ward (1963) and Burr (1970) point out clustering could be based on the minimum *sum* of squares within clusters resulting from each fusion rather than on minimal *increase* of this value. Such a procedure frequently leads to absurd results and is not recommended.

This clustering method may also be applied with dissimilarity measures other than D^2; these can be treated exactly as with D^2.

The technique has been proposed by several workers the first of whom was apparently Ward (1963) who described it as an "error sum of squares" strategy, then Anderson (1966) proposed it under the name of "minimum variance clustering." Orloci (1967a) also developed the strategy under the "sum of squares method," and finally Burr (1968, 1970) coined the term "incremental sum of squares." The last terminology seems the most descriptive of the four and so is to be preferred.

Since the strategy is capable of using mixed data and especially continuous variables, it is ecologically useful when the elements are defined in terms of such measures as cover abundance on the Domin scale. Using such results, Pritchard and Anderson (1971) found it useful for classifying a series of Scottish vegetation samples.

After standardizing his site data by converting them to proportions, Orloci (1967a) used the technique for classifying a sample of Canadian vegetation, and Hughes and Thomas (1971a) employed it for studies on Canadian marine benthos. The strategy is intensely clustering and shows the property of group size dependence; for this reason Lavarack (1972) found it unsuitable for his taxonomic work on orchids. In this it resembles the information gain strategy, and the mathematical relationships between these two strategies has been discussed by Burr (1970).

h. Minimal Variance

A method of clustering allied to that just described is one in which there is a minimal increase in the variance rather than the sum of squares within a cluster at each step in the fusion cycle. There are few published reports using this clustering cycle and its properties are not well known. Its algebraic formulation is given by Burr (1970) and it has been used for the study of orchids (Lavarack, 1972), earthworms (Wallace, 1972), and *Lomandra* (Sands, 1972).

i. Flexible

Lance and Williams (1966b), in commenting briefly on five of the strategies above, noted that there had been no previous attempt to generalize them into a single system with the result that separate programs had

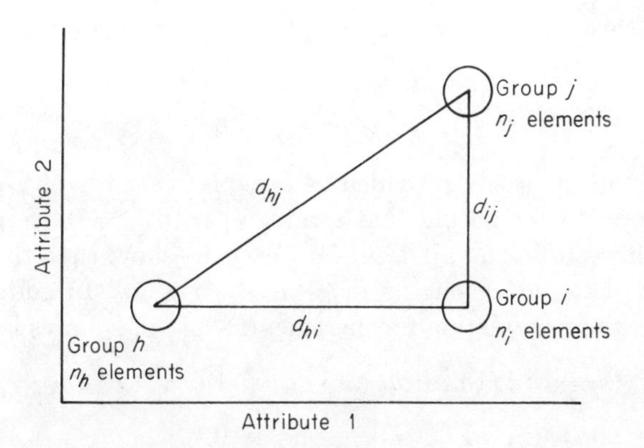

Fig. 8.6. The relationships between three groups of individuals with respect to two attributes. Distances d_{hi}, d_{hj}, and d_{ij} are measured between the centroids of the groups and each group contains n_i, n_j, and n_h individuals as specified.

to be written for each one. The general system they proposed is given in diagrammatic form in Fig. 8.6 for a distance measure. Any of the dissimilarity measures, and not only Euclidean distance (D, D^2), can be employed.

Lance and Williams (1966b, p. 218) said:

> "We first assume three groups (h), (i) and (j), containing n_h n_i and n_j elements respectively and with inter-group distances already defined as d_{hi}, d_{nj}, d_{ij}. We further assume that the smallest of all distances still to be considered is d_{ij}, so that (i) and (j) fuse to form a new group (k) with $n_k = (n_i + n_j)$ elements. The problem is solved if we can always express d_{hk} in terms of any or all of the quantities already defined; they are all, of necessity, known at the moment of fusion."

These authors continue: "We assume the linear relation $d_{hk} = \alpha_i d_{hi} + \alpha_j d_{hj} + \beta d_{ij} + \gamma \mid d_{hi} - d_{hj} \mid$ where the parameters α_i, α_j, β and γ determine the nature of the sorting strategy. It can be shown that the five strategies already mentioned are obtained when the parameters take the following values:

Nearest-neighbour: $\alpha_i = \alpha_j = +\frac{1}{2}$; $\beta = 0$; $\gamma = -\frac{1}{2}$

Furthest-neighbour: $\alpha_i = \alpha_j = +\frac{1}{2}$; $\beta = 0$; $\gamma = +\frac{1}{2}$

Median: $\alpha_i = \alpha_j = +\frac{1}{2}$; $\beta = -\frac{1}{4}$; $\gamma = 0$

Group-average: $\alpha_i = n_i/n_k$; $\alpha_j = n_j/n_k$; $\beta = \gamma = 0$

Centroid (this applies only to Euclidean distance squared):

$$\alpha_i = n_i/n_k; \qquad \alpha_j = n_j/n_k$$

$$\beta = -\alpha_i\alpha_j; \qquad \gamma = 0$$

All can therefore be easily provided as optional variants of a single computer program." After noting that strategies with $\gamma = 0$ are not always monotonic they state: "... it is in fact easy to show that the system is necessarily monotonic so long as $(\alpha_i + \alpha_j + \beta) \leqslant 1$." In addition to the above listing, the formulation for incremental sum of squares is as follows:

$$\alpha_i = (n_h + n_j)/(n_h + n_k); \qquad \alpha_j = (n_h + n_j)/(n_h + n_k)$$

$$\beta = -n_h(n_h + n_k); \qquad \gamma = 0$$

In a later paper Lance and Williams (1967a) proposed a new strategy described as "flexible" and based on their earlier general equation, and with the following constraints:

$$\alpha_i + \alpha_j + \beta = 1; \qquad \alpha_i = \alpha_j; \qquad \beta < 1; \qquad \text{and} \qquad \gamma = 0$$

The reason for the constraint of β becomes obvious by the example they consider when different values are applied. Williams *et al.* (1971a) show clearly that positive values of β give space-contracting, a zero value space-conserving, and negative values space-dilating strategies. Clearly clustering intensity can be varied, and Williams (1971) has termed β the "cluster-intensity coefficient." In practice it has become conventional to set β at -0.25 for the preliminary investigation of classificatory problems.

Flexible sorting ($\beta = -0.25$) has been used successfully in ecology on marine benthic data by Stephenson *et al.* (1970, 1972, 1974), Stephenson and Williams (1971), Williams *et al.* (1973), and Williams and Stephenson (1973). It has also been used for the classification of *Acacia harpophylla* (Coaldrake, 1971) and *Salvia* (El-Gazzar *et al.*, 1968; El-Gazzar and Watson, 1970).

In its usual employment with β fixed at -0.25, it could be argued that, in pragmatic use, the system is no longer flexible. With $\beta = -0.25$ there is a family resemblance between flexible in which

$$d_{hk} = 0.625(d_{hi} + d_{hj}) - 0.25d_{ij}$$

and median in which

$$d_{hk} = 0.5(d_{hi} + d_{hj}) - 0.25d_{ij}$$

Besides the above clustering strategies there are a variety of others not in common use. For details of these reference should be made to Sneath and Sokal (1973) and Fisher and van Ness (1971).

E. NONHIERARCHICAL CLUSTERING, CLUMPING, GRAPHS, AND MINIMUM SPANNING TREES

Besides the hierarchical classificatory procedures described above there are a number of nonhierarchical strategies. The theoretical bases of the majority of these are quite unrelated either to one another or those of the hierarchical strategies. For most of these nonhierarchical strategies there are too few published accounts of their application to ecological or taxonomic problems for them to be as critically assessed as was possible for the hierarchical strategies.

Graphs and minimum spanning trees have been included amongst these miscellaneous classificatory procedures largely because they appear to have closer affinities with classificatory means of reducing data than they have with ordination methods of data reduction. The application of such graph theory methods to classificatory studies is of recent origin and it is anticipated these methods will receive considerable attention in the future.

1. Density Clustering

In most of the strategies previously described for forming clusters, these depend on an initial fusion or division which thereafter is fixed for the remainder of the analyses. One exception to this applies to the divisive polythetic classifications in which entities may be reallocated if they appear misclassified by an early dichotomy (REMUL). The nonprobabilistic clustering now to be outlined is also divisive, but accepts that the composition of groups changes continuously throughout the earlier stages of the analysis. It begins with centers to which allocations are made—these may be entities or even positions in space, moreover these centers may be chosen arbitrarily or may depend on prior knowledge of the data.

Clearly before analysis is undertaken a decision must be made on how many centers are established and thus how many clusters are required. If only two centers are used this would result in a polythetic divisive dichotomy and by successive application a polythetic divisive hierarchical classification could be obtained.

Assuming n individuals to have been chosen as nuclei, all remaining entities are consigned to their nearest nucleus. When all have been allotted, the coordinates of the centroid of each group are calculated and the groups disbanded to allow all the individuals to be reallotted to the nearest centroid. Repetition of the process of clustering and reclustering continues until the system is stable, that is the constitution of groups is constant over at least two clustering cycles.

The technique does not appear to have been employed by ecologists but has been applied to the classification of the pea genus *Phyllota* (Jancey, 1966a,b). It is known that different starting points may give different clumps with the same data and so the method can be criticized for its possible ambiguity. Nevertheless it appears to us to show considerable promise.

2. Probabilistic Clustering

In contrast with the nonprobabilistic clustering method just described there is the direct probabilistic approach devised by Goodall (1966a) and which has been applied both the classification of grasses (Clifford and Goodall, 1967) and bacteria (Goodall, 1966b). It takes into account the relative frequencies of different attribute states in the array of entities, and on the bases of these, determines for each species pair the probability of obtaining a pair at least as alike as these. That pair for which the probability is smallest becomes the starting pair. This pair forms the nucleus of a group that share in common the property that they are statistically homogeneous at a given level of significance. When no further members may be added to the group without exceeding this level of significance it is removed from the original population and the process is repeated on the remaining set of entities. It should be noted that the technique gives weight to more rare attribute states and that the relative frequencies of these states change during the analysis with the segregation of each group of entities. The method is computationally very slow and so far as we are aware the clustering at each stage has to be undertaken by the user, both of which factors may account for the neglect of the method.

3. Clumping

The essence of the present clumping is that entities may occur in more than one clump; hence, we are concerned with overlapping classifications. Though this concept has found considerable favor in some fields of classificatory endeavor, for example, libraries (Needham, 1962), it has been little employed by either taxonomists or ecologists.

For the taxonomist the underlying theory of evolution suggests his classifications should be hierarchical and nonoverlapping. While this practice is strictly adhered to by the majority of taxonomists, an exception is to be found in the work of the grass taxonomist Bor (1960) who classified several individual species as belonging either to the genus *Saccharum* or *Erianthus* and included them under both genera.

The ecologist is not constrained by the evolutionary background of the taxonomist and here clumping is permitted, and may even be desirable. When classifying a sample of complex rain forest (Williams *et al.*, 1969), it was found useful to generate clumps by associating together individuals with a fixed number of their topographically nearest-neighbors (see also Williams *et al.*, 1973). In this way the influence of different densities of vegetation were minimized in a manner that did not seem possible with quadrat data. Having obtained the clumps they were then treated as quadrat data and subjected to hierarchical clustering strategies and ordination.

A similar clumping procedure was adopted for studies of marine benthos by Stephenson *et al.* (1970) and Stephenson and Williams (1971), who clumped sites rather than individuals. Such a procedure is the equivalent of obtaining a two-dimensional running mean and emphasizes the presence of rarer individuals in the sample. As with the rain forest data, the clumps are available for subsequent clustering or ordination.

With the above procedures, clumps have been produced by means of a single extrinsic attribute—proximity. They exploit none of the intrinsic data in their selection. It would generally be desirable to effect clumping by intrinsic attributes.

One recent study that fulfills this requirement is due to Yarranton *et al.* (1972) who classified a set of plant communities of the cracks of a limestone pavement. Here they grouped together species on the basis of the magnitude of the correlation coefficient between them and then used these magnitudes to generate clumps. A single individual of known identity was selected and all those strongly correlated with it were united into a single clump. This clump was then dissolved, a new individual chosen as the basis of a clump, and the procedure repeated. The addition of elements to a clump ceased when the correlation coefficients between the starting member of the clump and the remaining individuals were less than a given value. It should be stressed that with clumping, the total set of data is needed to establish the clumps prior to possible classification.

However, the groupings resulting from combining an analysis and its inverse into a two-way table are not clumps, though this terminology was employed albeit in inverted commas by Webb *et al.* (1969) in a study on the classification of rain forest when discussing concentration analysis as a means of data reduction.

Likewise in clinal taxonomic situations the same population may be in two separate clines but this does not cause it to be a clump but merely a group recognized in terms of two separate sets of criteria.

4. Graphs

Until recently (Gower and Ross, 1969), little attention has been given to graphs (as distinct from trees and dendrograms) as classificatory devices. As employed here, the term graph refers to a set of points, the relationships between which are expressed in terms of lines connecting the points, the lengths of the lines themselves being of no consequence. In numerical taxonomic studies the points represent entities and the lines given levels of similarity between the entities. Entities are only joined by lines when the similarities between them equals or exceeds a given critical value.

The manner in which graph theory may be employed to shed light on a classificatory problem will be considered first with reference to a hypothetical example and then with respect to two underlying theoretical concepts. Consider the similarity matrix of Table 8.1 in which the similarities between six entities are expressed as percentages.

If membership of a group implied all entities to be 100% alike with this example each entity would constitute a single group. When this stringent requirement for group membership is reduced to 90% similarity between entities there are four groups—A and B; C and E; with D and F as single membered groups. These results together with those for two lower values of similarity as the critical values for group membership are expressed graphically in Fig. 8.7. From the figure it can be seen that as the level of similarity required between entities is lowered for group membership, the number of groups that result is reduced. Furthermore, it is clear from Fig. 8.7 that two features of the graphs are of general interest. One is the magnitude of the difference between the groups at any stage in the analysis, the other is the extent to which the groups are internally homogeneous.

TABLE 8.1

SIMILARITY MATRIX (TRELLIS DIAGRAM) BETWEEN 6 HYPOTHETICAL ENTITIES[a]

	A	B	C	D	E	F
A						
B	90					
C	30	20				
D	80	40	10			
E	75	60	90	20		
F	60	70	80	35	85	

[a] Similarities given as percentages.

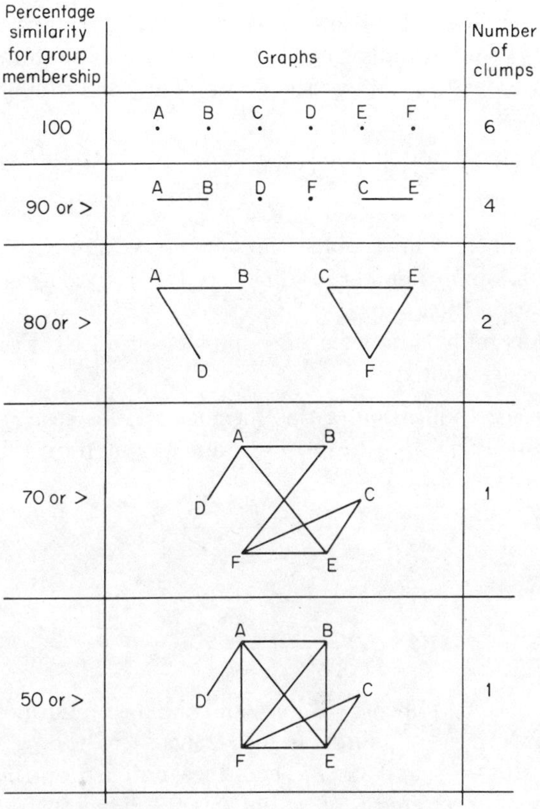

Percentage similarity for group membership	Graphs	Number of clumps
100	A B C D E F	6
90 or >	A B D F C E	4
80 or >	A B C E D F	2
70 or >	A B D C F E	1
50 or >	A B D C F E	1

Fig. 8.7. Graphical classification of the entities whose similarities are given in Table 8.1.

The magnitude of the difference between two groups is defined as the amount by which the similarity between entities must be reduced in order that two groups become united, and is referred to as the moat between groups. Thus at the 4-group level in the above example the similarities between entities must drop to 80% before entity D unites with group AB and so the moat between these two groups would be (90–80) or 10 units of similarity.

In that entities become united with those they most closely resemble at each stage of the analysis, this method of grouping has much in common with nearest-neighbor sorting but differs in that here individual entities may be joined to more than one other member.

The internal homogeneity or connectedness of a group is estimated in terms of the number of lines linking entities within groups relative to the number of lines possible, allowing that a certain minimum number of lines is required in order to establish the group. The connectedness of a group (C) is defined as:

$$C = \frac{X - Y}{T - Y}$$

where X = number of lines connecting entities within the group
Y = minimum number of lines required to connect all entities within the group
T = maximum number of lines possible if all entities in the group are connected

The manner of application of this formula may be seen with respect to Diagram 8.1 involving 4 entities S, T, U, and V united by 4 lines at a given level of similarity

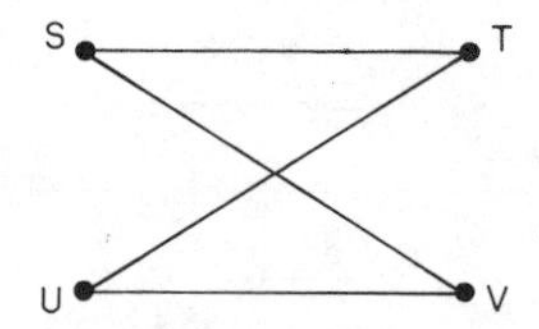

Any 4 points may be joined by 3 lines and the maximum possible number of lines connecting 4 points is 6. Hence, in the above example $C = (4 - 3)/(6 - 3) = 0.33$. It should be noted that the sequence in which the points are united is of no importance. Thus the three minimally connected graphs in Diagram 8.2 are all equivalent for this purpose.

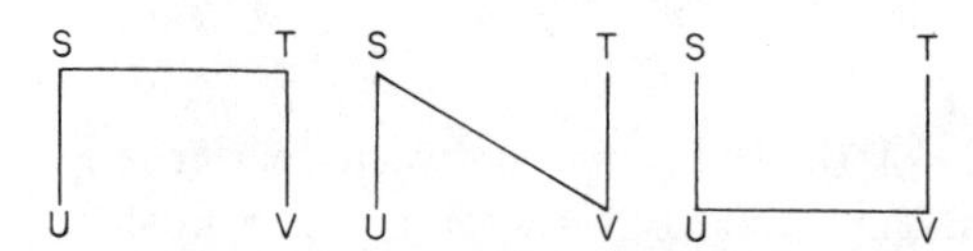

In general for N entities the minimum number of lines to connect all entities is $N - 1$ and the maximum number is $N/2(N - 1)$. Applying these formulas to the graph at the 70% level in Fig. 8.7 the connectedness is:

$$C = \frac{7 - 5}{15} = 0.13$$

whereas at the 50% level

$$C = \frac{9 - 5}{15} = 0.27$$

With this grouping technique the number of groups that may be recognized varies from unity for the whole population to the number of entities to be grouped. Any desired number of groups between these two values may be achieved by fixing the similarity index for linking entities at the appropriate value. Any similarity index may be employed. This method of clumping has been widely applied in taxonomic studies (grasses, Clifford, 1965; orchids, Wirth *et al.*, 1966; orchids, Lavarack, 1972).

An alternative approach is found in the work of Fager and McGowan (1963) in marine ecology. Having estimated the degree of similarity between all species pairs under consideration using their similarity coefficient, groups were formed by linking together species in such a manner that within groups all interpair similarities were equal to or exceeded a certain stated value. That is, every group had a maximum connectedness and there was no overlapping between groups.

5. Minimum Spanning Trees

An alternative method of grouping based on graph theory but (unlike the method above) retaining a measure of interentity difference is the minimum spanning tree. Here a set of lines joins pairs of points in such a way that no closed loops occur and all points are linked so that it is possible as it were to travel from any point to any other along one or more lines. The distances between points (dissimilarities) are chosen so that the total length of the lines is a minimum.

The determination of the minimum spanning tree may be undertaken in a number of ways (Gower and Ross, 1969). One algorithm is to begin with a given entity and join it to its neighbor which will be the shortest distance away; the next shortest distance in the system determines the next pair of entities to be associated and so on until all entities are linked, care being taken to avoid links which will generate loops in the line. Clearly there can be no more than $N - 1$ links in the tree where N is the number of entities.

The minimum spanning tree for the entities in Table 8.2 is given in Diagram 8.3.

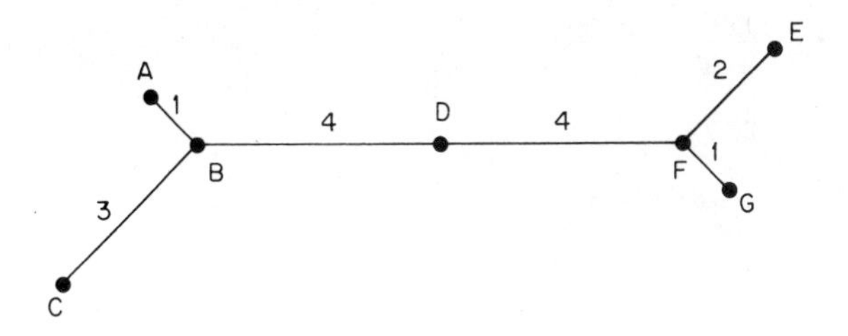

TABLE 8.2

THE DISSIMILARITIES BETWEEN 7 ENTITIES MEASURED IN TERMS OF THE NUMBER OF DIFFERENCES BETWEEN THEM

	A	B	C	D	E	F	G
A							
B	1						
C	3	2					
D	5	4	4				
E	3	4	4	4			
F	3	4	4	4	2		
G	2	3	3	5	3	1	

The lengths of the lines in the tree are the numbers of differences observed between the entities. Any dissimilarity measure may be employed.

As might be expected from its manner of construction the minimum spanning tree is closely related to the nearest-neighbor dendrogram and the latter may be constructed given the lengths of the lines of the entities they link in the tree. This may be confirmed by reference to the minimum spanning tree and Table 8.2.

With minimum spanning tree, the immediate relationships of the entities are expressed and so provide a check on ordinations (see Chapter 13) where entities clearly separated in multidimensional space be projected as close together in space of fewer dimensions. In effect, the minimum spanning tree may be likened to a backbone running through multidimensional space linking together entities with the shortest possible length of bone. Having so linked them, the whole is reduced to two dimensions by rotating the various portions using the points or entities as pivots.

9

The Handling and Interpretation of the Results of Computer Classifications

A. GENERAL COMPARISON AND INTERPRETATION OF RESULTS

It should be abundantly clear that a great range of classificatory techniques is available and that choices are required at many stages. These include (i) the form of the data, for example, whether binary or quantitative data should be collected and also whether data will be of more than one kind, (ii) for quantitative data whether there should be transformations and if so of what type, (iii) whether or not there should be data reduction and if so how and to what level, (iv) what type of similarity/dissimilarity measure should be chosen, (v) whether data should be standardized and if so how, and (vi) what type of classificatory strategy should be employed.

The choice of methods is often related to the nature of the data. For example, if mixed data with a high proportion of binaries is involved (as with much taxonomic work), the use of an information gain strategy such as MULTBET may be indicated. Ecological data with many zero entries and a few outstandingly high entries may well suggest transformation, use of the Bray-Curtis measure, and flexible or group average sorting.

The final choice of methods depends on how the data can be usefully interpreted by the biological user, a viewpoint which has been expressed strongly by Williams (1972) and by Greig-Smith (1971). However, it has given rise to criticisms that methods are being selected on a subjective basis to prove what it is the observer wants to prove anyway. Let us be honest and admit that there is an element of truth in these criticisms. Let us also attempt to define what it is we look for in the "best" computer classification.

In taxonomy an experienced worker can generally detect an entity which appears to have been misclassified. He can also judge whether a classification is broadly "better" or "worse." He makes these judgments on a basis of experience and intuition which may not be easy to quantify or even verbalize. In general, the evolution of satisfactory taxonomic classifications by numerical methods has been toward those that satisfy taxonomists and the satisfactory programs are tending toward the same sort of value judgments that are made by human brains. Having established such methods they can be applied to "unseen" taxonomic situations, where there can not be a prediction of the precise answers (otherwise why bother at all), but reasonable certainty that they will produce "something sensible." Furthermore, the prediction value of a classification may be used to test its merit. In general the "better" classifications have a high predictive value in that the accumulation of further attributes results in little change to the system.

Trial has shown, however, that the "best" taxonomic methods are not always "best" for ecological problems. In ecology, the situation is from some aspects much more fluid and complex than in taxonomy and from others much more amenable to impartial checking. The fluidity and complexity arise because there is no established hierarchy of either site groups or species groups to which (it is hoped) the analytical results might conform. The impartiality is because there are usually extrinsic data which have not been used in the analysis which can be used for checking purposes. An obvious example of this is topographical patterning. If no topographical data has been used in the analyses (and this is typical), then the method producing a coherent grouping of sites is likely to be preferable to one which produces incoherence and fragmentation.

In ecological analyses there are, in fact, two different kinds of criteria which can be used in comparing the effectiveness of classification. The first depend on the extent to which, using only intrinsic data, like is grouped with like, and the second depend on the extent to which groupings obtained by intrinsic data reflect extrinsic attributes.

Considering first the intrinsic situation, the most effective way of overseeing the results of both a normal and inverse classification is to present the data as a full two-way coincidence table. This is the original matrix resorted into the groups as indicated by the classification; it is convenient to indicate major and increasingly minor groups in this table, and to take the groupings to finer levels than are likely finally to be required.

Because massive amounts of data are usually involved in computer classifications, the dimensions of two-way tables preclude the possibility of their publication in a full form. A full table from a small data matrix is shown in Table 9.1. It is from Stephenson and Williams (1971, p. 23), and the data used were from topographical clumping of sites.

Perusal of two-way tables is desirable for two reasons, first, the location of "misclassifications," and second, to make decisions on the levels of classification to adopt. With respect to "misclassifications," as indicated earlier they can occur in divisive programs because entities otherwise similar may have been misdirected by an early dichotomy. Meanwhile in agglomerative programs they can occur because fusions begin where group affinities are weakest.

Misclassifications are easiest to locate when binary data are used, for example, when one site in a site group records several species as absent while in the other sites of the group they are present. When meristic or continuous data are used misclassifications are revealed by such things as a series of low values where high ones might be expected. In more succinct terms it is sometimes possible to improve constancy ratings (considering data as binary) or dominance ratings (meristic or continuous data) by transferring entities or attributes from one to another group. In later pages these ratings are discussed in more detail.

Resorting of matrices in this way relates to the work of Popham and Ellis (1971, p. 70) who stated: "A Zurich-Montpellier analysis (Kückler, 1967) was used for the combined R- and Q-type procedure, i.e. grouping species which co-occurred in samples, and grouping samples with re-occuring species, respectively. The analysis was by visual comparison of presence-absence data arranged in a series of increasingly more organized tables. The Zurich-Montpellier procedure may have been applied previously to sedentary marine faunas, but we have been unable to find published references to it in this context." The difference between the above approach and for example that of Stephenson and Williams (1971) is that the latter involved a minor visual correction to coincidence tables obtained by a different method.

The difficulties of locating misclassified entities are less formidable than might appear. In addition to purely visual methods, one can frequently find an entity which disturbs a pattern, for example, lowers the constancy in more than one attribute. Often this can be reallocated merely by inspection of the matrix. If not it can be either treated as a single aberrant entity or can be reallocated by simple calculation. For the latter the dissimilarity of the entity to each of the group means is obtained, using the same dissimilarity measure as previously. The entity is then allocated to the group giving the smallest dissimilarity. (See also Chapter 9, Section C.)

Perusal of two-way tables is also valuable for comparing classifications

TABLE 9.1

Two-way Coincidence Table Showing Site Groups (with Site Numbers) and Species Groups (with Species Numbers) (Benthos, Sek, New Guinea)[a]

Species groups	Species	Site groups: 1								2					3						4					5							
		55	56	57	58	59	60	61	62	25	26	27	30	37	48	49	50	52	53	54	1	2	3	12	13	4	5	6	7	8	9	10	11
I	1	1	1	1				1	1	6	4				6	3	2	1	1	1	6	6	6	3	3	6	5	4	5	5	5	3	1
	2									2	1										5	5	5	9	9	3	2	4	3	3	4	4	3
II	8									51	52	52	52	52																			
III	3																									1		1		1	1	1	1
	13							1	1																								
	14																																
	20											2																					
IV	15	1													1	2	2	1	2	1													

	5	6	8	8	16	16	16	10	10																						
V	6	1	2	2	4	4	4	3	3																						
	11	2	2	2	2	2	2																								
	4									1		2	2	2	1	1		1													
	10												1	1								1	1								
	12									1	1	1	1	1				1				1	1	1		1		1	1	1	2
VI	16									1																					
	17									1	2		2	2																	
	18									1			1	1																	
	19									1	1	1	1	1								1	1							1	1
VII	7																		1	1	1	1	1	4	3	3	3	4	4	3	2
VIII	9																							1	1	2	1	2	1	1	1

[a] Data use "proximity values" obtained by clumping adjacent sites. For dendrogram of site groups see Fig. 9.1; for ordination of these groups see Fig. 9.2. (From Stephenson and Williams, 1971.)

(*Continued*)

TABLE 9.1[a]—Continued

Species groups	Species	6A		6B			7			8			9																		
		47	51	29	35	36	23	24	28	14	15	16	17	18	19	20	21	22	31	32	33	34	38	39	40	41	42	43	44	45	46
I	1	11	10	5	5	2	20	18	16	5	14	26	43	43	44	42	23	42	48	46	37	26	33	46	48	32	36	36	36	35	50
	2	1	1	1	1		8	6	6	17	21	30	23	23	21	15	14	15	15	4	3	4	7	3	6		3	3	3	4	4
II	8			2		2	1	2	2									1				1									
III	3						2	2	2				1	1	1	1	1	3	1	3	3	4	1	4	4	3	3	3	3	3	4
	13						1	1	1	1			2	2	2	2	2	3	2			1									
	14												1	1	1	1	1		1	2	2	2			2	2					
	20				1															1	2	1		1	1	2					1
IV	15																														

	5																	1	1	1		1			1						
V	6																														
	11												1	1	1	1	1	1	1	1	1			1	1	1					1
	4	2	2	1	2			1	1									1	1	1		2	1	2		1	1	1	2	2
	10			1	1	1				1	1																				
	12	1	1	1		1		1	1		1	1																			
VI	16	1	1	1	1		1	1	1																				1	1
	17			2		2		1	1																						
	18			2	2	1	1	1	1																						
	19			1		1		1	1	1	1																				
VII	7									1	1																				
VIII	9																1		1	1		1									

[a] Data use "proximity values" obtained by clumping adjacent sites. For dendrogram of site groups see Fig. 9.1; for ordination of these groups see Fig. 9.2. (From Stephenson and Williams, 1971.)

in terms of intrinsic attributes. The simplest method is to use cell densities in the tables. These are expressed as percentages which can be arbitrarily graded, for example, Stephenson *et al.* (1972) used very high (VH) for 100–95%, high (H) 94–66%, medium (M) 65–33%, and low (L) 32–0%. It will be clear that a classification in which there are more very high and low densities effects a "crisper" classification than one in which the entries tend to run "across the board" giving more medium densities.

Cell densities depend on the number of infrequently occurring species which are used in the analysis, clearly with a larger number there will be more blank entries and a lower density. Until methods of species reduction can be standardized between problems, comparison of results of different surveys or of different classificatory strategies by means of cell densities will be misleading. The more ubiquitous species will make major contributions to cell densities, but are likely to appear in many cells. (In Table 9.1, species group I illustrates this situation.) They have high constancy (often repeated) but low fidelity. Species of intermediate ubiquity are (hopefully) concentrated in certain cells, and hence show high constancy and high fidelity. (In Table 9.1 species group IV illustrates this situation.)

Constancy and fidelity depend on the number of site groups which are generated. By increasing the number of site groups, the ubiquitous species are likely to completely fill more of the cells, and the density of these cells will increase. Meanwhile, because more cells are likely to be occupied, at least by ubiquitous species, fidelity will decrease. In part, the number of site groups which are required depends on an ecological decision on whether constancy or fidelity is to be favored. If the former site groups are likely to be characterized by constancy (hopefully high) of different combinations of many species groups. This gives its own particular "flavor" to community concepts; one which has been outlined in dealing with marine bottom communities (Stephenson *et al.*, 1972).

Characterization by fidelity is likely to be greater if there are fewer species groups, and also if fewer of the less ubiquitous species have been eliminated during data reduction. Again it is an ecological decision whether the communities should be recognized by the less common species. To the logic of some ecologists, both animal and plant, this has been a reasonable approach to classification. In marine benthic ecology on the other hand, the stress has been toward constancy at the expense of fidelity.

In noting that an increased number of site groups increases constancy and decreases fidelity, we should also note the comparable effects of increasing the number of species groups. It will be rare for any two species to show exactly equivalent distributional patterns within surveys and both constancy and fidelity of species groups in site groups will increase with the number of species groups.

Using the same original data, and generating the same numbers of site groups and species groups, respectively, it ought to be possible to effect direct comparisons between two classificatory methods by comparing the two levels of cell densities, for example, by noting which gives the greater number of cells of very high and high densities. In practice there are difficulties. For example, in an attempt by Stephenson *et al.* (1972) to compare the discreteness of classification using group average and flexible sortings, it was found (as might be expected) that the weakly clustering group average strategy produced many small site groups. To make anything like an effective comparison it was necessary to take the group average strategy to the 30 site group level, discard 12 site groups with only one or two sites in them and make the comparison between 18 "large" group average groupings and 20 flexible sorting groupings.

The existence of small groupings, particularly small site groups raises further questions as to whether the apparently constant species in them owe something to random occurrence. For example, in a site group of two sites, the minimal number of recordings (one) gives 50% constancy. As such it may mean very little, and in these types of terms many of the groupings obtained by weak sorting strategies may be justifiable on one premise and be quite suspect on another. We shall later discuss tests of significance but meanwhile an alert common sense is required.

Using constancy as a measure of the effectiveness of classification, instead of working on cell densities which are depressed by infrequent species, it is preferable to work via the individual species. (These are also unaffected by the number of species groups which are selected.)

As indicated earlier, decisions are required on the number of site groups required. The most convenient method appears to be to generate a slight excess of the number considered desirable. When a hierarchical classification has been used, the dendrogram indicates the higher fusions, and each can be explored in turn to determine how cell densities, constancy of species, and fidelity of species is altered by fusion.

When the classification is or has become nonhierarchical, perusal of the two-way table appears to give almost as good a guide as the hierarchically produced dendrograms as to where fusions of site groups might be made. If this fails, another method is available. To show affinities in such cases it may be desirable to ordinate the site groups (Chapter 13). If there has only been recourse to dendrograms, ordination should be used with the hierarchical methods which involve sharpened clustering. This is because of the phenomenon, already considered (See Chapter 8, Section D, 3a) of group size dependence. A large group becomes increasingly difficult to join and separates from the hierarchy at a level determined partly by its relative affinities with other groups and partly by the number of sites it contains. The differences between affinities as deduced by a dendrogram and those obtained by ordination are illustrated by Fig. 9.1 and 9.2 drawn from a study of marine benthos in New Guinea (Stephenson and Williams, 1971). It will be evident that site group 9 has become spuriously isolated merely because it contains many sites.

Possibly because of this phenomenon it is not always desirable to truncate each branch of a dendrogram at the same level. The levels adopted should be those at which the two-way coincidence table reveals the optimum amount of ecological sense. Though much differential truncation involves subjective decisions, it can be argued that objectivity is maintained in that the "cut-off" values are explicitly stated.

To date we have dealt primarily with the normal (site) classification. The easiest routine for handling the similar problems of the inverse classification appears to involve handling the site group situation first, and then to examine, via the two-way table, the effects of fusion of species groups.

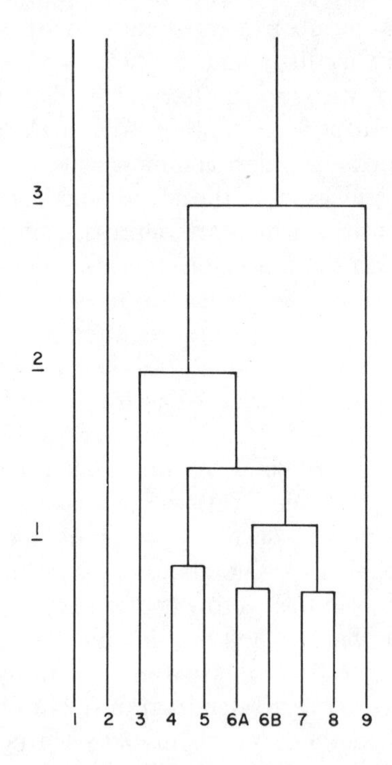

Fig. 9.1. Portion of a dendrogram of a set of site groups (benthos, Sek, New Guinea). Ordinate represents fusion values obtained from the Bray-Curtis measure; sorting is flexible. Note the isolation of site group 9. See also Table 9.1. (From Stephenson and Williams, 1971.)

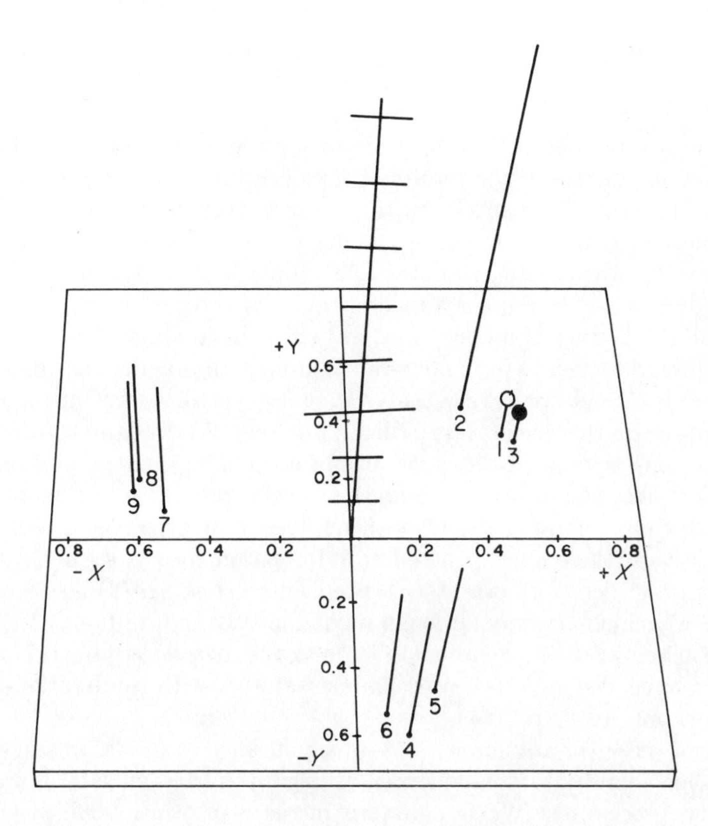

Fig. 9.2. Three-dimensional ordination of the same set of site groups as shown in Fig. 9.1. Viewing is from the "south-west" with about 45° angle of depression. The central aerial represents the *z*-axis with the zero-axis marked with a broader bar. The white ball on group 1 indicates a major positive deviation (> 0.66) in the fourth dimension and the black ball on group 3, a major negative deviation (> 0.66) in the fifth dimension. Note the closeness of site group 9 to groups 7 and 8. (From Stephenson and Williams, 1971.)

In a classification using binary data, the results of classification can only be examined on the bases of cell densities, constancies, and fidelities. If data are quantitative further criteria are available, for example, numerical abundance or biomass; these we may collectively designate as "dominance." To express dominance in readily understandable terms it is convenient first to obtain the means of values in the different cells of the coincidence table (using either species means or species group means) and second to convert these continuous values into a number of arbitrary grades. Examples will be found in Stephenson *et al.* (1972).

B. APPLICATION OF SIGNIFICANCE TESTS

Various attempts have been made to replace with tests of significance personal judgments of the results of classification. These are most simply applied to each of the species in turn. For a given species, recordings are split into several subsets corresponding to the different site groups and the data are then tested to determine whether or not the recordings are at random. Possibly there is a theoretical quibble regarding this procedure in that the testing is applied to data which have already been through procedures designed to form nonrandom groups. In practice the application of the tests tends to reveal that even after classification an important proportion of the species are still at random. Work is proceeding (W. Stephenson) to explore the possibility of comparing these proportions as a test to replace common sense judgments in several phases of classification.

Meanwhile regarding the tests themselves, four types have been used. Respectively, these are (1) based on information theory, see Field (1969, 1971); (2)χ^2, see Kikkawa (1968), Stephenson *et al.* (1972) also used this test in a preliminary way although relying mostly on intuition; (3) F-test, see Williams and Stephenson (1973) and Stephenson *et al.* (1974); and (4) we have also undertaken preliminary studies with the Kruskal-Wallis test (Siegel, 1956, p. 184).

Use of these (or any other) tests of significance of species in site groups has some immediate consequences as regards forms of data and classificatory procedures. When data are meristic or continuous instead of binary the chances of showing significant differences are greatly increased. For example, consider the tabulated meristic data which follows.

	Sites in site group 1					Sites in site group 2				
	1	3	5	7	9	2	4	6	8	10
Species 1	60	58	65	62	70	1	2	1	2	1

By any test the values in site group 1 are significantly greater than those in site group 2. However, when converted to binary (i.e., presence/absence) form, the two subsets are identical.

Some classificatory strategies, for example, group average, tend to produce single-site site groups and in these clearly the within-group variance is undefined. This precludes the use of certain tests of significance and decreases greatly the chances of showing significant differences. For the effective use of significance tests it is desirable that site groups should be of approximately equivalent size. This biases the choice of sorting strategies toward those with group size dependence, for example, flexible. If a space-conserving strategy such as group average is adopted it is desirable to accept groups at widely desparate levels on the dendrogram to obtain approximately equivalent sizes of groups. If the embarrassment of single-site sitegroups still remains, these may be reallocated to one of the other groups.

Of the tests outlined possibly the F-test is open to the greatest criticism, because there is an underlying assumption that the within-group values are normally distributed: Clearly this does not apply when a species is completely missing from a site group. Because of the problems associated with normality of distribution, possibly a nonparametric test such as the Kruskal-Wallis is indicated.

One value of tests of significance of species in site groups is in relation to data reduction; this we have considered briefly earlier (see Chapter 7, Section B). We will assume judgment was used to select an arbitrary cut-off level (generally in terms of ubiquity or abundance) below which uncommon species were excluded. The site groups, derived from the originally accepted species, are used to test how many of these are conforming. In one study (Stephenson *et al.*, 1974), it was shown that most of the originally accepted but nonconforming species were close to the cut-off level (this was in terms of variance).

Species originally excluded can also be tested and a few, again near the cut-off level, were shown to conform. In general the judgments of exclusion levels were justified, but in detail it appears that a postanalytical testing is desirable.

Conformity tests are being used here as a data reduction device, and open the way for a repeat classification using the species conforming to the original grouping. Contrary to expectation the site groups obtained are not always identical with the originals. A further classification is possible using species above the original exclusion level which do *not* conform to the original site pattern. Hopes that an alternative pattern would be found, possibly suggesting control by other abiotic variables, have not materialized in the trials made to date (Stephenson *et al.*, 1974).

Further applications of conformity tests are envisaged along the following lines:

(a) The number of species which conform to site groupings at different levels of the site classification dendrogram. It is expected that the optimal levels will correspond to those selected visually.

(b) Comparisons between classifications employing different dissimilarity coefficients and sorting strategies to determine which give the greatest numbers of conforming species.

(c) By taking the site groups derived from biota attributes, and replacing these by abiotic attributes to investigate conformity of extrinsic attributes. We consider extrinsic attributes in more detail in Chapter 11.

Meanwhile it should be noted that the idea of a conforming species may require interpretation in terms of the more familiar concepts of dominance, constancy, and fidelity. When nonbinary data are used it appears that most of the conforming species are characterized by marked differences of dominance-cum-constancy in the different site groups.

C. COMBINATION OF STRATEGIES

We know of virtually no cases in which identical results are obtained when different numerical classificatory methods have been used on a mass of data of sufficient volume and complexity to justify a numerical approach. By the methods outlined above it is sometimes possible to discard a method completely because the results appear nonsensical, but in others the choice of which is "best" can not readily be made. In such circumstances we can have reasonable confidence in those sections of the analyses giving agreement between two methods, and are left with a doubtful residue. Most of this can be fitted reasonably easily into the groups already accepted by a process of reallocation. Stephenson *et al.* (1972) state: "The method for stations was to define, for any reference group, a group vector whose jth element was the mean of the transformed counts or weights for that species in that group. This vector could then be compared with a disputed station by means of whichever measure (Bray-Curtis or Canberra metric) had been used for the initial classification."

This combination method produced sharper sorting than either of the methods which would, individually, have been regarded as satisfactory. It did this at the expense of losing some of the advantages of hierarchical classification, and also at the expense of tedious noncomputer comparisons of the composition of alternative groupings. These comparisons involve subjective decisions; for example, the minimal number of stations (sites) extracted from two long and broadly dissimilar lists which could be regarded as the same.

D. DENDROGRAMS

It has become conventional with hierarchical clustering strategies to present the results in the form of a dendrogram. In terms of graph theory, a dendrogram is a rooted tree the nodes of which link together the entities or clusters being classified. The root of the tree may be at the top or bottom depending on such factors as whether the classification strategy is divisive or agglomerative and whether relationships between entities have been estimated as similarities or dissimilarities. The successive nodes represent the clusters resulting from fusion or separation of prior clusters and the interval between nodes represents an increase in some measure designed to show the consequence of fusing the clusters. Thus, there may be an increase in information content following fusion of the clusters, an increase in the sum of squares within clusters of the Euclidean distance between clusters. Further the increase may be in terms of the nearest- or furthest-neighbors within the clusters, to name a few of the measures available.

The clusters arising from different similarity measures and clustering strategies applied to data from the same set of elements will very likely differ according to the procedures adopted. It is therefore of interest to be able to compare dendrograms in a quantitative manner so as to be able to determine accurately the extent of such differences. The manner in which comparisons may be made will be considered with reference to three dendrograms resulting from the classification of a set of seven entities for each of which a single meristic attribute was available. The entities and their attributes scores follow:

Entity:	*a*	*b*	*c*	*d*	*e*	*f*	*g*
Attribute score:	2	6	9	14	20	21	23

The dendrograms resulting from classifying the seven entities using nearest-neighbor, furthest-neighbor, and group average clustering strategies and Euclidean distance as a dissimilarity measure are given in Fig. 9.3.

From this it is clear that, as expected, the entities in each analysis clustered into two groups. One measure of dissimilarity between dendrograms is to determine for each dendrogram the number of nodes separating each species pair. For example, in the nearest-neighbor dendrogram three nodes separate b and d while in the furthest-neighbor case four nodes separate them. Let the number of nodes separating two entities be m_1 for dendrogram 1 and m_2 for dendrogram 2. The differences $| m_1 - m_2 |$ are summed over all pairs of species to give the value D (not to be confused with Euclidean distance), which is then a measure of the difference between the dendrograms. The value D may be scaled by reference to the total number of comparisons possible thereby permitting dendrograms with different numbers of entities to be compared.

An alternative method of comparison of dendrograms is to consider the

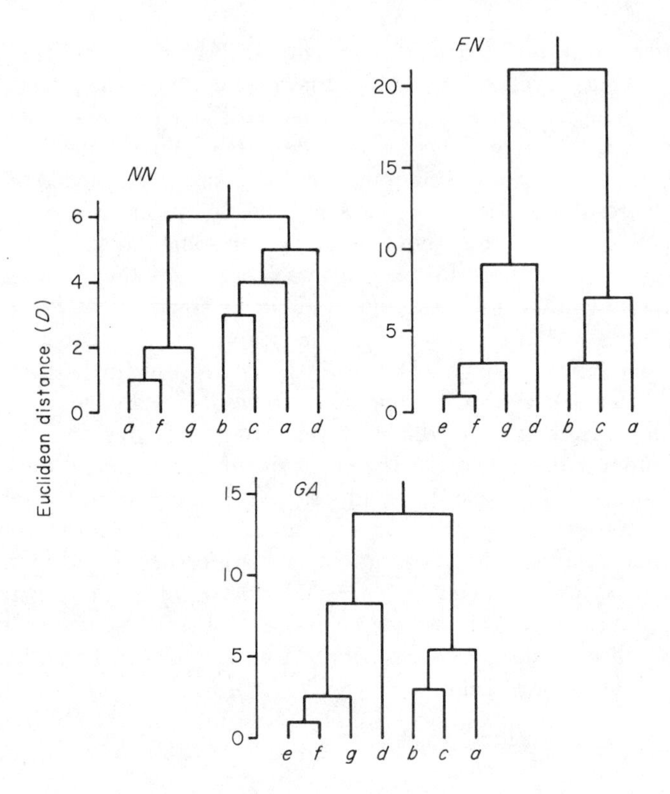

Fig. 9.3. Dendrograms resulting from clustering the same set of entities using three different clustering techniques: *NN*, nearest-neighbor, *FN*, furthest-neighbor, and *GA*, group-average clustering.

composition of groups rather than the disposition of the nodes. The value of either m_1 or m_2 for a pair of entities in the same group will be zero. When estimating D it is easy to count the number of cases for which m_1 or m_2 is zero in one dendrogram but not the other. The total number of such cases is denoted by G and is a measure of the number of transfers required to convert one dendrogram into the other. Like D, G can be scaled with reference to the total number of possible comparisons. The smaller the values of D and G the more similar the dendrogram under consideration with respect to the structure and group composition.

As an example of the usage of these two indices reference may be made to Fig. 9.3 in which the dendrograms have been truncated at a value of 2.6 on the vertical scale to provide clusters for demonstrating the method of calculating D and G. Five groups result for both nearest-neighbor and group average clustering when truncated at this level. Furthermore the composition of these five groups is identical for both dendrograms though their structures are different. That is, for this pair of dendrograms G but not D is zero.

The group compositions for both the NN and GA dendrograms truncated at the 2.6 level are tabulated below.

Group 1	e, f, g
2	a
3	b
4	c
5	d

As defined G is the number of instances in which the number of nodes separating any two entities is zero in the first but not the second classification plus the number of instances in which the number of nodes separating any two entities is zero in the second but not the first classification. Here with e, f, and g in the same group in both classifications and the remaining groups being singletons in both, there is no instance in which the number of nodes separating two entities is zero in one but not in the other classification. Hence, G is zero.

Turning to the calculation of D it follows that $D\ (e, f) = |0 - 0| = 0$ since e and f are in the same group in each classification. However, because the structure of the dendrograms is not the same value of D is not always zero. For instance $D\ (e, d) = |4 - 3| = 1$. Summed over all entities $D = 6$. The results of comparing all three dendrogram in terms of their structure (D) and group composition (G) are given in Table 9.2.

TABLE 9.2

ABSOLUTE VALUES OF D AND G FOR THE DENDROGRAMS PRESENTED IN FIG. 9.3

	G		
D	NN	GA	FN
NN	×	0	2
GA	6	×	2
FN	14	14	×

From the table it is clear that with respect to both structure and group composition nearest-neighbor clustering more closely resembles group average than furthest-neighbor clustering. This is of considerable interest for in their nontruncated forms the furthest-neighbor and group average dendrograms are identical in both structure and group composition.

However, with a truncation at (Euclidean distance), $D = 2.6$ the furthest neighbor dendrogram produces six groups and the group average dendrogram only five groups. This is because with furthest-neighbor sorting entity g unites with e and f at a level of $D = 2.5$. The production of extra groups at a given similarity level reflects the highly intensely clustering nature of furthest-neighbor sorting already discussed above when considering clustering strategies.

The measures D and G were devised by Williams and Clifford (1971) and are somewhat similar to those earlier proposed by Sokal and Rohlf (1962). However they have an advantage that they are an invariant property of the hierarchy and independent of its method of presentation. The cophenetic correlation coefficient proposed by Sokal and Rohlf was calculated as follows. Each dendrogram was divided into slices by drawing across it a series of equally spaced lines parallel to the base, that is, at right angles to the vertical. Each zone was numbered serially, there usually being about 10 zones. For each pair of entities the zone in which the node uniting them was noted in each pair of dendrograms under consideration. The resultant pairs of values were recorded for each pair of entities and the product moment correlation coefficient between them noted. The greater the value of r the more similar are the dendrograms.

10

Difficulties in Numerical Classification

Most of these are considered in an ecological context, because here the situations are intrinsically more complex and the difficulties are more formidable. Many are illustrated by reference to studies of the "bottom communities" of benthic marine invertebrates, and they range across the entire field from the sampling program and nature of the data to problems over the type of computer used.

A. OBJECTIVES IN CLASSIFICATION

The object of classification of site/species data is to produce a number of tolerably discrete groups or patterns of co-occurrences. Data are collected with reference to a particular biota, for example, in macrobenthic surveys it is usual to employ one of the types of dredge or grab, but in microbenthic work one of the types of corers (see Holme, 1949, 1951, 1953, 1955, 1964). The distance between samples is related to the size of samples taken and is smaller with corer samples. It is not surprising that the scale of patterns obtained in microbenthic surveys is much finer than that obtained in macrobenthic work. The existence of small-scale microbenthic patterns within the area of a single macrobenthic association is a clear indication of heterogeneity of a different order within the latter. This suggests that it will be a completely impossible undertaking to define a macrobenthic community in terms of occupying an homogeneous environment. In more general terms we can only define a community in biotic terms by the organisms within that community in which we have declared an interest, and in abiotic terms by the abiotic factors in which we have declared an interest. Of the two it appears that the organisms have the greater claim.

We have assumed in the foregoing that we are interested in one or more communities, and have further assumed that we know what we mean by a "community." One doubts whether this second assumption holds other than in a small minority of cases and this is discussed by Stephenson (1973a,b). It is sometimes easier to infer the real interest of an investigator from the design of his sampling program then from anything else. For example, if a study is made over a short period of time, approximating to a single *augenblick*, it is a fair assumption that spatial relationships are of interest but temporal ones are not. In this context one may question the value of all "Environmental Impact Statements" based on a single set of observations within a short time span. All sampling programs have a spatial restraint for purely logistic reasons and this implies selection of a traverse, an area or volume within which the sampling will occur.

We shall consider traverses first. These are the simplest (one dimensional) form of sampling. Use of a traverse generally implies that there is suspected to be a marked change in biota or in abiotic factors along a selected line, and hence assumes previous reconnaissance. In benthic work traverses usually run at right angles to depth contours, and this is tantamount to a declaration of interest in depth. The usual requirements are then to know at what changes in depth are there noteworthy changes of biota, and which biotic groups occupy which depth ranges.

There are three main ways in which stations can be distributed along a traverse; they can be deliberately spaced on features of interest (e.g., in the center of each biotic zone), they can be at random, or they can be at equal intervals. Random location is often advocated for statistical reasons, but is uneconomical and also is unsuited for determining discontinuities since such sampling is designed to offset random fluctuation in the habitat. With regard to the economics suppose there are three zones and that only three samples can be taken. The chances that each zone will be sampled are quite low; alternatively, to be sure of sampling each zone a large number of samples would be needed. With respect to discontinuities, in the usual analyses, sites are grouped by their biotic similarities and one would expect in advance their similarities would be greater between sites in close proximity. Classification of randomly spaced sites would mostly be trivial in reflecting their random proximities; there would be too close a relationship between topographical distance and ecological distance. The more sensible approach is equal spacing—in a depth context this could be either in a vertical or horizontal line.

Areas can be studied either by spaced traverses or by a grid pattern. The former implies an expectation of a greater effect along one axis than another and may raise problems unless traverses are of equal length. In general grid patterns seem preferable, particularly in early surveys. Sampling of volumes is a particular feature of planktonic work, where an interest in depth exists. The solution adopted is of vertical traverses to some sensible (i.e., nonrandom) pattern of covering the area of interest.

Let us consider area sampling in more detail. Here there may be three objectives. The first is to subdivide the total area into a series of subareas, and since the sampling is presumed to have been for biota, the subarea will be biotically delineated. As indicated earlier it may be desirable to transform the data for this type of classification. The second objective is to group the biota into a series of species groups each of which will characterize certain areas or combinations of areas. For this type of classification, it appears desirable to standardize the data. A third objective is to know which species (and groups of species) live in which ranges of environmental conditions. It appears that this last objective may be more important to more investigators than generally realized. In the usual analyses, it is attained by linking species to site groups and site groups to abiotic variables, but there are other and more direct approaches to this problem, as outlined later. These can lead to species groupings of a different kind from those above, and to comparative descriptions of the habitats of the different species in abiotic terms.

In an unknown area, assuming sampling to a grid pattern, the first question will be how far apart the samples should be spaced. This can only be approached by actual sampling of the area. A sensible approach is to take two samples at n km apart, and by simple calculation (e.g., Bray-Curtis dissimilarity measure, raw data) or by visual estimation arrive at some concept of the degree of similarity/dissimilarity between the two. If they are too similar, take a third sample $2n$ km from one of the others, and continue in geometric progression ($4n$, $8n$, etc.) until a "reasonable" difference has been obtained. If the two original samples are too different, take a third sample $n/2$ km from one of the others, and again continue in geometric progression ($n/4$, $n/8$, etc.) until a "reasonable" similarity has been obtained.

After establishment of a grid pattern, further problems are likely to arise; for example, there may be large areas with an homogeneous biota and small areas in which large changes in biota occur over short distances. The latter may require a smaller scale grid interpolated within the original.

Turning to volume sampling, this presents particular problems since a three-dimensional equally spaced lattice is unlikely to be practicable. Equality of spacing in the most important axis (e.g., depth) seems an initial requirement. Whereas area sampling (at a given time) gives two-dimensional data, the data from volume sampling are at least three-dimensional with respect to space. This greatly complicates the analytical procedures, as outlined later.

We next briefly consider the importance of time in relation to sampling. It is doubtful if any biotic situations are completely stable for the duration of the interest of an investigator. To simplify matters it is sometimes possible to neglect time, but if there is some doubt as to the propriety of this, an exploratory investigation on the lines of the grid dimensions should be undertaken. Samples are taken with a time interval of n units, and subsequently at $2n$, $4n$, etc., or at $n/2$, $n/4$, etc., units of time.

Part of the problem of sampling in time is that at least three time scales are involved: diurnal, seasonal, and annual. All are known to be important in plankton and this plus the three-dimensional nature of the "instantaneous" data creates analytical problems which do not appear to have been solved. In benthos, seasonal and annual changes are known to occur, but the analytical methods for handling the data are only recent (Williams and Stephenson, 1973; Stephenson *et al.*, 1974). These we shall consider separately at a later stage.

B. CHOICE OF DATA

This refers to the form in which the data on the species are recorded under site headings. The simplest form is binary (presence/absence) but for reasons given variously elsewhere, this should be avoided as being the least informative. Also for reasons given earlier graded and ranked data should be regarded as second best. This leaves as practical choices data upon numbers or weights or activities.

Temporarily disregarding the last, it appears certain from reanalyses of Petersen's (1914) data by Stephenson *et al.* (1972) that it is essential to treat numbers data and weight data separately: There appears to be no way of having a "number-plus-weights" measure. Of the two, numbers are more easily obtained and most workers appear to agree with Field and Macfarlane (1968) that the extra labor of weighing should not be undertaken unless justifiable. Also numbers are basically in meristic form while weights are continuous, and this can facilitate analyses. Finally there has been more analytical experience of handling numbers data; in particular little consideration has been given to the appropriate transformations to apply to weights data.

However weights are preferable to numbers in cases where species interactions or productivities may be of parallel or later interest. If weights are to be used it is theoretically desirable that they be as biomass dry weights excluding inert material. The decision to gather such data might well be influenced by time and motion conditions.

In phytosociological studies a wider range of parameters (e.g., percentage cover) is available, and Greig-Smith (1964) should be consulted.

While separate from the general themes of the present work, one should note (see Chapter 3, Section B,2), that site grouping can be carried out effectively by the use of attributes other than species occurrences or abundances. In soft bottom ecology such factors as the particulate nature of the sediments, their calcium carbonate, spaces of interstitial pores, and pH and *Eh* values at different horizons could be employed. The list could be widened to include temperature, depth, surface illumination, water turbidity, and any if not all of the standard oceanographic and sedimentary parameters. The problems of this type of classification are numerous, and particularly concern the independent weighting and standardizing of attributes. For example *Eh*, oxygen concentrations and sulfide concentrations are not independent variables and inclusion of all might give a special weighting to one fundamental attribute as does size to correlated variables in taxonomy. Also attributes are measured in different scales, frequently with inbuilt transformations, for example, a logarithmic one with pH. Standardizations involving centering are necessary.

C. CHOICE OF STRATEGY

It will be evident that Sokal and Sneath's (1963) hopes that the subjective elements would be removed from taxonomy by the introduction of numerical methods has not materialized. There is a multiplicity of measures of similarity, of methods of standardizing data, and of sorting strategies. In ecology the range of choices is even more bewildering because of data reduction and transformation possibilities. Many of the techniques which have been used have a distinctly arbitrary flavor, and have been severely criticized, particularly by British workers (see Jardine and Sibson, 1971a).

There are two approaches to the evolution of a "good" classificatory strategy. The first, the mathematical approach, is exemplified by Jardine and Sibson (1971a), who have listed desirable criteria. In practice, as Williams (1972), has shown acceptance of these criteria would effectively restrict choice to the nearest-neighbor fusion strategy—possibly the least useful to the biologist.

The second approach is essentially that of "customer satisfaction." A biologist confronted with a number of alternative methods is likely to try a variety, and find that he has preferences. This may be because in his analysis, without consciously realizing it, he wishes to stress constancy—

the method which does this is then preferred. With increasing knowledge or experience of what stresses and biases are built into each of the various methods he may be able to eliminate or reduce the methods which are compared. Also with increasing exposure to numerical methods the ecological objectives themselves tend to become more clarified and examples have been given earlier in this chapter.

Williams (1971) has stated that many of the strategies in current biological use ". . . . simulate rather closely the mental processes of the professional taxonomist or ecologist. It ensures that, given an intractably extensive set of data, the computer can provide a solution that will reasonably approximate to that which the professional would obtain intuitively, had he sufficient time to study the data closely; such a solution will therefore provide an invaluable starting-point . . . ." One might add that the methods not only simulate, they also stimulate. In addition to assisting in clarifying the data sets, they help in clarifying mental processes. It is a two-way interaction, and the biological problems stimulate the development of new numerical methods.

D. PRESENTATION OF RESULTS

The entire rationale of numerical analysis is that more data can be handled than by intuitive analysis, and in the context of delineating site groups by species as attributes, this results in each group being characterized by a potentially large group of species. Each of these has different levels of constancy, fidelity, and dominance and, hence, a description which includes the totality of data becomes too complex for comprehension. Two forms of compression appear to be inevitable. The first is to concentrate only on that limited number of species which shows the greatest diagnostic differences between site groups. Hence, many of the species used in the analysis do not appear in the final verbal presentation and the end result resembles those of the intuitive types of analysis. The difference is that in the latter most species are removed before the analysis, whereas in numerical analyses they are removed at the final stage. A priori one might expect the numerical methods to give "better" results because they involve consideration of a much greater proportion of the total data.

The second form of compression is because values of dominance, constancy, and fidelity often obtained by numerical analyses are continuous and capable of expression to several "significant" figures. To assist interpretations it seems necessary to impose grading systems on them. Again it is preferable to use the more complex data during the analysis (meristic or continuous) but as a concession to human comprehension it has to be simplified to a graded form for presentation.

E. THE TIME FACTOR IN ECOLOGICAL ANALYSES: MULTIDIMENSIONAL DATA

A major series of problems arise around the time factor in ecological survey. Many definitions of communities and associations exclude the time element either deliberately or by implication. Thus Möbius (1877) defines a biocoenosis as a community whose total of species is mutually linked under the influence of the *average external conditions of life,* and Hesse (1924) redefines it as the grouping together of the organisms living in a uniform part of the habitat and which in the selection and requirements of the species correspond with *average prevailing conditions.* (The italics are not in the originals.) Such definitions presuppose one can determine average conditions and apparently exclude from the community those species which occur only during non-average circumstances. A more pragmatic approach would be to restrict the biocoenosis to the species permanently present and this Petersen (1914) allegedly did in his study of bottom communities by restricting his characterizing species to non-seasonal species.

Other definitions involve dynamic concepts, for example, Resvoy's (1924) definition of a biocoenosis as a population system occurring under given ecological conditions and maintaining itself in dynamic balance. Such ideas led to Emerson's (1939) concept of a superorganismal community.

On balance it seems desirable to include the time factor if at all possible. In earlier benthic studies by W. Stephenson this was not possible either because of the complexities of the situation (Moreton Bay work) or because of external constraints on the sampling program (New Guinea work). The complexities when time is included are formidable and here again we can contrast the classificatory problems of the taxonomist and the ecologist. The former deals with a near-static situation in which variations with time are at the subspecific and population genetics level and are amenable to standard statistical treatment. The ecologist begins by having, at least in certain tropical environments, a situation at a given "instant" at least as intractable as any taxonomic problem. Added to this his attributes (species in sites) are changing either on a slow time base (coral reefs) or on a seasonal (bottom communities) or on an hourly (trawled invertebrates) basis. As stated earlier, the taxonomic equivalent would be characters with unheard of chronological flexibility.

Some of the problems of one direct approach to the problem are illustrated by the work of Hailstone (1972) discussed by Stephenson *et al.* (1972). Hailstone surveyed the sublittoral fauna at the mouth of the Brisbane River and each month of the year his data revealed site groups and species groups related through two-way coincidence tables. However from month to month because *both* kinds of groups altered their composition, it was difficult to find a basis for comparison.

Williams and Stephenson (1973) considered the problems of analysis in this type of situation. They pointed out that when an area is sampled at appropriate time intervals, the data matrix is three-dimensional; if there are q quadrats or sites, t times, and s species, the dimensions are $(q \times t \times s)$. (It should be noted that these dimensions refer to the matrix and at this stage have no relevance to whether the quadrats are in one, two, or three dimensions of topographical space.) The above authors refer to the usual method of analysis which is to regard the matrix as having two dimensions $(qt \times s)$, and state that the main information extracted is whether the groupings obtained by classification reflect primarily spatial or temporal differences or both. Beyond the primary dichotomy there are almost invariably, groupings of samples containing different quadrats at one time and different times at one quadrat. Such results approach the meaningless.

Williams and Stephenson (1973) suggested partitioning of the matrix, producing for example site/species classifications eliminating time and time/species classifications eliminating sites. They used an analysis of variance model. Considered first in a site/species context, they showed the dissimilarity between two sites could be expressed as an estimate of variance due to species, and that [after suitable scaling by t (s-1)] it was identical with the squared unstandardized Euclidean distance between sites (over all species) after each site had been centered to the station mean. Using this dissimilarity measure, a site classification using species as attributes can be made by any of the sorting strategies. In the study referred to above, Stephenson and Williams used flexible sorting with a Euclidean distance measure of dissimilarity based on data transformed to cube roots.

In a similar way the dissimilarity between two species can be expressed as an estimate of variance due to sites (eliminating times). After suitable scaling it is identical with the squared Euclidean distance between species (over all sites) after each species has been centered to the species mean.

Problems arise with two-way tables. If the values used for classification (i.e., centered values) are used, for every site species entry there will be two values to contend with, one with site centering and one with species centering. In the above paper, the two values were summed but in a later extension by Stephenson *et al.* (1974) they were handled separately.

By partitioning it is possible to obtain a two-way table for site/species classifications, another for those involving times/species, and a third for times/sites—the last being concerned with total population changes. Further, it is possible to establish the relative heterogeneity in the data associated with each of the pairs of dimensions, and this is conveniently expressed as mean variance associated with a single comparison. The authors show that by using the q and s data (eliminating t), the mean variance attributable to species for a single intersite comparison is the same as the mean variance attributable to sites for a single intersite comparison, and further, that the interaction value between any two of the three dimensions is the same. Being able to give relative "importance" values to the matrices of site/species, times/species, sites/times, and interactions is an important adjunct of the method.

Another is that it provides a simplified form of data reduction using the same model as applied in classification. Within a given site, the site mean of all species is obtained and this is repeated for all sites. A given species is then considered, the numbers (summed over all times) in each site is obtained and the site means are then subtracted in each site. For each pair of sites, the squares of the differences for the given species is obtained; these are then summated over all pairs of sites. This gives a measure of the importance of each species as regards the measure used for classification. Species can then be arranged in hierarchical order and data reduction can be based on a single arbitrary decision as to where to break the string.

The general approach and the corollaries of the paper by Williams and Stephenson (1973) have been too recent for all their implications to be appreciated. The use of the same model for data reduction as for classification could be applied to most dissimilarity measures with additive properties, for example, the Canberra metric. Consider the matrix tabulated below.

	Sites				
Species	1	2	3	4	5
1	50	90	100	200	150
2	0	1	0	1	0
3	0	0	1	0	0

The classificatory importance of species 1 would be the summation of all site pair Canberra metric measures involving species 1, i.e.,

$$\frac{90-50}{90+50}+\frac{100-50}{100+50}+\frac{200-50}{200+50}+\frac{150-50}{150+50}+\frac{100-90}{100+90}+\frac{200-90}{200+90}$$

$$+\frac{150-90}{150+90}+\frac{200-100}{200+100}+\frac{150-100}{150+100}+\frac{200-150}{200+150}=\text{ca. }3$$

Meanwhile the classificatory importance of species 2 would be

$$\frac{1-0}{1+0}+\frac{1-0}{1+0}+\frac{1-0}{1+0}+\frac{1-1}{1+1}+\frac{1-0}{1+0}+\frac{1-0}{1+0}+\frac{1-0}{1+0}=7$$

and of species 3

$$\frac{1-0}{1+0}+\frac{1-0}{1+0}+\frac{1-0}{1+0}+\frac{1-0}{1+0}=4$$

These values show that with the "raw" Canberra metric, species with an alternation of low and zero measures are given the greatest importance, and show the lack of emphasis given to numerical importance.

The Williams and Stephenson method was applied to fuller results by Stephenson *et al.* (1974), whose data consisted of species in sites sampled at three monthly intervals over a period of two years. The results of the times/species classification showed that subordinate groups produced confusion comparable to that indicated earlier in $qt \times s$ analysis. The problems were resolved by taking the $t \times s$ dimensions of the matrix and regarding time as two dimensional. From the partitioning, matrices of seasons/species and years/species were obtained and there was a marked increase in conceptual clarity.

Stephenson *et al.* (1974) also used tests of conformity of species to site groups and to time groups (for tests of conformity see Chapter 9, Section B). In the present context two conclusions were important. The first was that the classificatory importance of a species was not indicated in a complete fashion by its place in the variance contribution hierarchy. The second conclusion was that species with a similar pattern of conformity to site patterns (for example) were not concentrated into the same species groups. A revised method of species classification was clearly required, and the combination of two-way tables by the original method was criticized.

A simpler method of classification after summation is readily conceived. Reverting to the $q \times t \times s$ matrix, we can summate species over times and obtain a $q \times s$ matrix and again summate species over sites and obtain a $t \times s$ matrix. Each can be classified by any of a number of methods, and there may be advantages in avoiding the Euclidean distance measure because of its extreme dependence on the transformation employed. Raphael and Stephenson (1972) used a Bray-Curtis measure with a square root transformation for site groupings and time groupings, but have incorporated a standardization by totals prior to obtaining the corresponding species groupings. Raw values have been used in the final two-way tables. The steps in the process and the final results are possibly more amenable to straightforward ecological interpretation than when the Williams and Stephenson (1973) with the Stephenson *et al.* (1974) modifications were used. A rough estimate of the relative importance of sites and times with respect to species can be obtained by comparing the numbers of species which conform to site groupings as compared to time groupings, although this is complicated by the fact that different numbers of groups, and hence different numbers of degrees of freedom are likely to be involved in the two sets of tests.

F. MACHINE DEPENDENCY

Most of the classificatory programs described above are highly complex, not in concept, but in terms of the manner in which they are designed to maximize efficient usage of the computer for which they are written. In practice it is usually difficult to transfer a program from one machine to another for a variety of reasons, some of which will now be considered.

The usage, even of the same program language is likely to vary from one to another computer centre so that characters legal at one place may be illegal at another. A problem related to the usage of language is the nature of the compiler which may translate an imported program differently from its machine translation at source. Then to conserve space some programs have been written directly in machine language and so are highly machine dependent.

The usual reason for writing in machine language is to conserve space in the memory of the machine and it is on the matter of memory that so many problems become complex. Even with the largest machines memory is likely to be exceeded even with modest classificatory programs demanding access to disc storage and recall.

To be versatile the program should have options on standardization of attributes, measures of similarity, and a range of clustering strategies. These requirements are included in most classificatory programs and when the length of them is considered it is not surprising that few have been written.

The average ecologist or taxonomist is not likely to possess the expertise to write his own program, and unless he is a highly competent programer he will be unable to adapt a program from elsewhere. Hence he is obliged to make use of existing programs and so is largely dependent on the institution possessing them for cooperation.

11

Relationships of Species to Extrinsic Factors in Ecological Analyses

As indicated earlier the objectives of ecological analyses can be manifold with primary interest either in site groups or in species groups. One of the reasons for the interest in site groups is that they are likely to show topographical patterns and if these can be elucidated in terms of abiotic contrasts, it may be possible to link the species groups to causal environmental factors.

If the real objective is to elucidate these links there are implications with respect to the general methods of classification such as have been outlined earlier, and also there are alternative and more direct methods which might be employed. Trials by Raphael and Stephenson (1972) have shown that, if the Bray-Curtis dissimilarity measure is used in preference to the Canberra metric, the site groups which are obtained are more homogeneous with respect to abiotic attributes. The implications of this appear to be that a "reasonable" stress on the dominant species is preferable to stress on the infrequently occurring ones if indications of the importance of abiotic factors are required. Contrary to expectation the best "indicator" species, at least in the above studies of marine benthos, were not the uncommon ones.

An earlier deliberate attempt to lay stress on extrinsic attributes was made by Tracey (1968) who classified a series of pastures which had been subjected to a variety of treatments: burning, mowing, etc. The plant species composition in each of a number of sites was recorded in binary (presence/absence) form and the treatment was similarly noted as applied/not applied. With the data being binary, association analyses could be and were applied. It will be remembered that these are divisive monothetic and use χ^2. Originally the data were divided at random into two, the first a prediction or construction set, and the second a validation set.

The construction set of sites was now taken and for each species present in them its association with the treatment was estimated using χ^2 as the association measure. Assume there to be 40 sites with a species where presences and absences are associated with whether the sites have been treated or otherwise as indicated in the two-way table below.

	Treatment		
Species	Applied	Not applied	
Present	21	1	22
Absent	4	14	18
	25	15	40

Here it is clear the species presence is closely associated with the site having been treated and its absence with the treatment not having been applied. Having calculated the values of χ^2 for all species from such tables the primary subdivision of the construction set of sites into two subsets is undertaken on the basis of the species/treatment association with the highest value of χ^2. One subset of sites will possess this species; the other subset will lack the species. Each of the subsets is now analysed in a similar way and further subdivisions are undertaken until none of the values of χ^2 obtained attains statistical significance, at which time the analysis into further subsets is regarded as complete.

The validation set of sites is then classified divisively into subsets using the *same species* as were used in subdividing the construction set and in the *same order*. There was of course no reference whatsoever to the extrinsic attribute (although it will be noted that the species chosen had been related to it in the construction set). The differences in species composition between the final subsets of the validation set are now tested for statistical significance. If the test is positive it may be assumed the extrinsic attribute is influencing the classification. Such a test would have been inappropriate to apply to differences between the subsets of the construction set for there the subsets were constructed in such a way as to maximize the influence of the external attribute.

In the validation set no attempt has been made to maximize the association between the species and the external attribute, which means that more often than not, the differences between the subsets are less marked in the validation than the construction set. The validation set therefore acts as a check against overestimating the importance of the external attribute.

It will be noted in Tracey's (1968) example that each of the extrinsic attributes was handled separately, and that data throughout was in binary form. Stephenson, Godfriaux, and Smith (unpublished) have developed a method of investigating the relationship between sampling attributes and recordings of species in which attributes are handled separately, but in which data are less constrained. The sampling attributes were dissected into either binary form (e.g., deep sites/shallow sites) or multistate (e.g., six sampling areas). The species recordings were in meristic form.

The sites are then grouped by a particular sampling attribute— for example, into the two depth ranges. The recordings of a given species are then tested to see if they occurred in a significantly nonrandom manner within the site groupings. The procedure involved one way analysis of variance, but since data are likely to be other than normally distributed, a nonparametric test (the Kruskal-Wallis) is employed. It permits immediate classification of the species by their relationship to a sampling attribute; for example, species are divided into those occurring most abundantly in deep water, or most abundantly in shallow water, together with a group for which significance has not been established.

The above method of handling sampling attributes can deal equally effectively with topographical and temporal subdivisions and can handle incomplete data. For complete data in which there is a special interest in the temporal situation, the variance partitioning method described earlier (see Chapter 10, Section E) seems preferable.

Extrinsic classifications such as described above have been little used in ecological studies and have apparently not been used in taxonomic studies. This is possibly because of their being monothetic divisive strategies applicable to larger data sets than are generally available to taxonomists. Without large data sets the divisive separation of subsets will cease after producing only a few subsets because of the significance test applied to χ^2.

For a more detailed account of divisive classification with respect to extrinsic attributes, reference should be made to Macnaughton-Smith (1965) who considers the handling of binary data without missing values, to Lance and Williams (1968b) who discuss the problems of handling mixed-data matrices including missing data and to Simon (1971) whose book is a detailed exposition of the topic.

12

Diversity and Classification

The study of ecological diversity has generated a very considerable literature effectively commencing with Hutchinson (1957) and MacArthur (1957). In many respects this literature appears isolated from that on numerical classification, and we now attempt a partial synthesis.

We first consider some of the measures of diversity and the interpretations which have been made from them, and then consider the relationship of diversity to classification.

A. MEASURES BASED ON NUMBERS OF SPECIES

In its initial use diversity meant nothing more than the total number of species in a sample or area and it has been used in this way by such ecologists as Gleason (1922), Patrick (1949), Hutchinson (1959), and Hessler and Sanders (1967). More recently numbers of individuals have been taken into account as well as the numbers of species.

Sanders (1968) elaborated and noted that because the number of species obtained varied with the sampling intensity, comparisons should be made between the curves obtained by plotting the number of species (S) against the number of individuals (N) in the collection. While Sanders refers to "diversity values," it is not clear how these are derived from his curves—he presumably refers to the rate of species increment in the lower parts of his curves. An alternative is to determine by interpolation the number of species to be expected per 100 collected individuals. The characteristic feature of Sanders' work was that he made no mathematical assumptions on relationships between numbers of species and numbers of individuals. His "rarefaction method" has been used by recent American workers on marine benthos including Boesch (1971) and Young and Rhodes (1971) who both used other methods in addition.

Sanders' work has been criticized by Hurlbert (1971) who notes that the interpolated values whereby Sanders estimates the number of species per 100 individuals (Sanders, Table 2) is overestimated and has errors (due to sampling) which range from 12 to 53%. Sanders' "rarefaction method" does appear to be least reliable in the early parts of his curves where the S/N values will be most dependent on chance. On purely mathematical grounds it would be preferable to concentrate on the later portions of his curves where the data are more massive. There are logistic problems in obtaining such a bulk of data.

Meanwhile other workers had made assumptions on the form of the S/N curves, based essentially on their obtaining straight line curves when a function of the number of species is plotted against the same or another function of the number of individuals. An early attempt, by Fisher *et al.* (1943) was based on a log/log relationship, with the diversity index α given by $S = \alpha \ln[1 + (N/\alpha)]$.

There has been increasing evidence, summarized by Edden (1971), that S plotted against $\log N$ gives a graph approximating to a normal distribution curve, and already Margalef (1951, 1957) has incorporated this into a diversity measure: $D = (S - 1)/\ln N$. This measure stands or falls on the correctness of the logarithmic transformation.

Hurlbert (1971) has noted that two measures exist based on numbers of species, and can be considered with minimal mathematical manipulation. He distinguishes between "numerical species richness" or just "species richness" which is the number of species contained within a specified number of individuals or specified biomass, and "real species richness" or "species density" which is the number of species in a standard collecting area (e.g., 1 km^2). This is a useful distinction, and of the alternatives it appears that the second has been most used, for example, by Woodwell (1967), Whittaker and Woodwell (1969), and Stephenson and Williams (1971). However, Odum (1967) worked from standard species richness, using 1000 individuals in each case.

B. MEASURES BASED ON THE PROPORTIONS OF SPECIES PRESENT

Most later workers and some earlier ones have considered that a diversity measure should include two components—the total number of species present and also the way in which these species make their numerical contributions. This is implicit or is stated by such workers as Simpson

(1949), Lloyd and Ghelardi (1964), Pielou (1966, 1969), Margalef (1968), Hurlbert (1971), and Edden (1971). Sanders (1968), following Whittaker (1965), has distinguished between "species diversity" based on the numbers of species present, and "dominance diversity" based on the proportionality of species, and this is a useful distinction to retain. The present measures are of dominance diversity.

One of these indices has recently been developed by Edden (1971) and like Margalef's (1951), it is based on the "lognormal law." It incorporates the standard deviation (σ') from the $\log N/S$ curve. The measure is $\log N - \frac{1}{2}(\sigma')^2$. The index has been little used and its properties are as yet little understood. It is interesting among the measures of dominance diversity in that it is probably less dominance biased than the remainder.

Apart from McIntosh's (1967a) measure outlined later, the remaining measures considered below are all concerned with probability, uncertainty, and information content.

The information content of a biotic assemblage in terms of species is equivalent to the uncertainty as to which species in the assemblage will be encountered in the next observation. If there are many species present and all have the same abundance it will be difficult to predict the next encounter (probability of prediction low), and there will be great uncertainty (high information content) and high complexity and diversity.

Pielou (1969) has given a lucid and mathematical account of the main probability and information theory indices of diversity. The earliest, due to Simpson (1949), is based on the probability that a second individual drawn from a collection will belong to the same species as the first—this probability is then the complement of the diversity. Allowing for the fact that a collection is a random sample of a complete population, the diversity D becomes

$$D = 1 - \frac{1}{N(N-1)} \sum_{j=1}^{S} n_j(n_j - 1)$$

where N is the number of individuals collected of all species, S is the number of species, and n_j the number of individuals of the jth species.

Hurlbert (1971), in a trenchant criticism of the concept of diversity measures, derives another measure based on the same essential principles as Simpson (1949). His measure of "probability of interspecific encounters" D^1 gives as a complement the following diversity measure:

$$D^1 = 1 - \frac{N}{N-1}\left[1 - \sum_{j=1}^{S}\left(\frac{n_j}{N}\right)^2\right]$$

which may be written as

$$D^1 = 1 - \frac{1}{N(N-1)} \cdot 2 \sum_{ij} n_i n_j \qquad 1 \leq i < j \leq S$$

where n_i and n_j are successively the products of the number of individuals of all possible species pairs.

Two information theory measures of diversity have been discussed earlier (Chapter 6, Section E; see Appendix for derivations). They are the Shannon index by our notation

$$H = N \log N - \sum_{j=1}^{S} n_j \log n_j$$

and the Brillouin

$$H(B) = \log N! - \sum_{j=1}^{S} \log n_j!$$

Both may be scaled to give diversity per specimen by dividing by N and scaled values can be derived to base 10 by convenient tables in Lloyd *et al.* (1968). The Shannon index has been more extensively used by ecologists generally as a dissimilarity measure than the Brillouin. For examples, see MacArthur (1955), MacArthur and MacArthur (1961), Lloyd and Ghelardi (1964), McIntosh (1967a), and MacArthur and Wilson (1967).

The various measures given above do not form a comprehensive list but illustrate the general principles which underlie them and indicate the diversity of diversity measures.

C. MEASURES OF EVENNESS OR EQUITABILITY

These relate directly to and are usually derived from diversity measures incorporating proportions of species. The higher the diversity (D) in an assemblage, the greater the tendency for equal numbers of all species—in ecological terms high evenness implies numerical codominance of many species, and low evenness implies marked dominance of a single species. It is easy to calculate the maximum diversity (D max) when all species present are equally abundant and the minimum diversity (D min) when all but one species are represented by single individuals and the one species contains the remainder.

Two conversions from diversity to evenness are

$$E^1 = \frac{D}{D_{max}} \quad \text{and} \quad E = \frac{D - D_{min}}{D_{max} - D_{min}}$$

(It may be noted that there is a resemblance between E and the formula used for determining the connectedness of a group; see graph theory, Chapter 8, Section E,4.)

A related measure can be used when the diversity measure is restrained between zero and unity. This is the complement of the diversity measure, i.e., $1 - D$, and this Patten (1962) has defined as a redundancy measure. As Hurlbert (1971) points out it can also be regarded as a measure of unevenness.

The E^1 index depends on species richness while the E values do not. Both indices depend on the sample size because more species are collected as the sample increases—hence, theoretically an indefinitely large sample is needed (Lloyd and Ghelardi, 1964; Pielou, 1967). Hurlbert (1971) quotes the following authors as having ignored this point: Patten (1962), Goulden (1966), Monk (1967), Barrett (1968), Pulliam *et al.* (1968), Buzas and Gibson (1969), and Sager and Hasler (1969). One way of overcoming this difficulty is to work from collections made in as completely uniform and identical a manner as possible.

An alternative measure of evenness has been proposed by Lloyd and Ghelardi (1964). Knowing that the Shannon diversity is due to given numbers of actual species, a calculation is then made of the number of hypothetical equitably distributed species which would give the same diversity.

An interesting index which has been called a diversity index, but which Pielou (1969) has shown is a uniformity index (U) is due to McIntosh (1967a). It is the Euclidean (ecological) distance of the assemblage from a point with zero individuals. Pielou (1969) has shown it can be converted to a diversity index by subtraction from the summated population of all species (N). It can be made independent of the size of N by conversion to a quotient viz., $(N - U)/(N + U)$.

D. IMPORTANCE OF DOMINANCE IN DIVERSITY MEASURES

Using Sanders' (1968) terminology, the diversity indices in common use are measures of dominance diversity and while supposed to include two components (species richness and proportionality of species), they in fact sacrifice the first. We have already noted this in considering the information theory indices where single records hardly contribute at all to diversity since log 1 = 0 to all bases; the same applies to McIntosh's (1967a) Euclidean distance model.

Dominance diversity has been criticized by various workers including Dickman (1968) and Sagar and Hasler (1969) because of its insensitivity to rare species, and by Margalef who stated (1968, p. 18): "... an area with greater diversity of butterflies would yield more species, and ... there would be a higher proportion of rare varieties." Because the interest in diversity measures is often in relation to tropical and subtropical environments, and because here there are many rare species, one might well question whether any of the existing indices measures in a satisfactory way the real focus of interest.

With a background of classificatory methodology it is evident that measures of species richness are based upon binary data (presence/absence), eliminate dominance, and have relationships to fidelity concepts, while the measures used for diversity appear overbiased toward dominance.

The existence of McIntosh's index and those based on information content show that diversity measures and the similarity/dissimilarity measures used in classification have some common ground. One suspects that there has been polarization of interest between the different fields and that workers interested in diversity are not familiar with the range of techniques used in classification. In classification, the various Manhattan metrics such as the Bray-Curtis and Canberra metric can be used to give a graduation from stress on dominancy (Bray-Curtis, untransformed data) to fidelity (Canberra metric without zero adjustment). Such measures can not be used as diversity measures—these require differential distances from an origin with zero individuals of zero species. An obvious possibility to reduce the importance of dominance in diversity measures is to use transformed data.

E. INTERPRETATIONS OF DIVERSITY

There seems to be general agreement that diversity measures should be applied within or between ecological units. For example, among recent authors Edden (1971) uses the term "ecological community" in her definition of a diversity index while Hendrickson and Ehrlich (1971, p. 1) say: "As most ecologists view the situation, species diversity is a measure of the relative richness of a community biota, or of the complexity of a segment of an ecosystem." It is quite clear if one worker uses the term "community" or "segment of an ecosystem" in a broader fashion than another, he is bound to "land-up" with a higher diversity measure.

This point has long been realized, for example, by MacArthur (1965) and Sanders (1968), who recognize that comparisons should be made on a within-habitat basis. In Sanders' case, his habitats were mostly "soft estuarine and marine oozes" (which were assumed each to be a uniform habitat) and sand. In the sand he obtained greater diversity and suggested this was due to a greater variety of microhabitats.

The main problem about a within-habitat restriction is to define and delimit a habitat. Benthic environments, including those dealt with by Sanders, are usually not capable of precise delimitation even in terms of the most "likely" environmental parameter—sediment type. We are usually a long way from the stage of knowing which habitat discontinuities to use to fairly encompass our diversity measures. The only feasible way of dividing the environment (by means of site groups) is to classify them using the species as attributes. We can then derive site groups characterized by many species in considerable abundance (high "dominance diversity," often accompanied by high "species richness"), by a single species in considerable abundance (low "dominance diversity" accompanied by either high or low "species richness"), or by negative criteria (low "dominance diversity" accompanied by either high or low "species richness").

Let us assume that we are comparing the diversities of two equal areas. Let us further assume that in the first area the sites form two groups (based on species as attributes) while in the second they form 20 groups. (We have further to assume that classifications were truncated in exactly comparable ways.) Then in the first area we must clearly expect a reduced apparent diversity. In essence we cannot say anything meaningful about diversities of areas until we know the scale of patterning of the areas.

In the one ecological field in which we know the literature in reasonable detail, that of the marine "communities" of soft bottoms, it is evident that there have been few or no intensive investigations of scales of patterning. As already indicated, there are indications that in inshore tropical and subtropical waters under some estuarine influence this is on a quite small scale (Stephenson and Williams, 1971; Stephenson *et al.*, 1974). The only other indirect evidence of pattern scales is in the difficulty that some workers have had in recognizing communities with clear dominants. The literature has been summarized by Stephenson *et al.* (1970), who have shown a general tendency for this difficulty to increase in warmer waters, and by Pearson (1970) who has shown that in cooler waters the difficulty is greater in inshore and varying environments. (The importance of varying environments we shall discuss later.) Meanwhile much of the discussion of diversity bears on the increased diversity of tropical environments, see, for example, Fischer (1960), Klopfer and MacArthur (1961), Pianka (1966), Sanders (1968), and Thorson (1952). At the level of species richness there is an immense literature in biogeography and in systematic, faunistic, and floristic treatments of various taxa. Surprisingly little of the literature deals with possible spatial "patchiness," the only references found to date are Pielou (1966) and Lloyd (1967). Both deal with it somewhat theoretically and without comparative measurements.

When comparisons are made between areas of different sizes, it is evident that, even if the scale of patterning is the same, one would expect to find more patterns ("associations" or "communities") in the larger areas. It appears that this factor has not always been adequately considered, although it is duly stressed by MacArthur (1965) in a "habitat" context.

Earlier we have shown that there are difficulties in applying classificatory methods to data cubes in which there is variation in species found in different sites and within different seasons, and have also noted the exclusion of seasonal ideas in many of the definitions of communities. If we accept the reality of seasonal associations or patterns of seasonal occurrence, many of the previous arguments about site patterns in terms of area again apply. One possibility is to restrict diversity comparisons to a given season, paralleling their previously suggested restriction to a given area. This would be completely unreal in many ecological situations, for example, marine zooplankton or phytoplankton in temperate seas show marked seasonal variations while in tropical seas there is much less by way of a gross seasonal pattern. By judicious choice of the season of comparison one could prove just about anything.

The only solution is to effect complete seasonal comparisons and hence compare the number of seasonal patterns (? "communities") at different situations. We already know or suspect that there will be obvious seasonal patterns in temperate latitudes and that there are likely to be ill-defined patterns, at least in organisms with short life cycles, in tropical situations. The elucidation of these tropical patterns of seasonality would be an important contribution to analyzing causes of diversity.

In recent work on the grab benthos of an area in southern Moreton Bay (Stephenson *et al.*, 1974), it was noted that the seasonal patterns in two successive years were far from identical and that there were area patterns, seasonal patterns, and annual patterns, each making their contribution to the total diversity. The concept of seasonal patterns, as regularly occurring phenomena contributing to diversity, does not conflict with the concepts of a considerable number of workers who equate high diversity with high stability of populations. The concept of different annual patterns, with no evidence of a regularity of reoccurrence, contributing to overall diversity is, however, contrary to the general views of the stability/diversity school. There is nothing intrinsically wrong with the argument that an environment with annual instability should be more diverse than one with annual stability. Indeed if diversity is equated to uncertainty, the more uncertain the environment the more uncertain the biota, and, almost inevitably the higher its diversity.

In discussing chronological stability and instability, clearly there are likely to be differences in scale. Gross instability leading to catastrophic reduction in biotas will clearly reduce diversity, whereas instability within "acceptable limits" may well increase it, as indicated above. The "acceptable limits" will be in terms of the range and availability of organisms capable of rapid invasion of changed areas, and what applies to a flooded near-estuarine situation after a season of excessive rainfall can not be expected to apply to an oceanic island devastated by a volcanic eruption.

It is clear that the annual changes in temperate and tropical environments have not been adequately compared over a sufficient range of environments for different annual patterns to be compared. It seems that such comparisons might be revealing.

We have not considered as such the different availabilities of niches in situations of different diversities—on this there is a formidable literature. From a niche viewpoint, the notion of seasonal communities, is merely delineation of seasonal niches with a possible distinction between "resident" and "seasonal" species. The change from year to year may equally effect distinctions between "perennial" from "transient" occupants. The fact remains that such distinctions appear to have been inadequately investigated; indeed even the classificatory techniques for their recognition are only now becoming available.

The development of programs to explore both within and between seasonal diversities offers exciting prospects for the taxonomic study of fossils. Here the bedded nature of the deposits in which the fossils are preserved provide a time scale and so could be sampled at different depths for comparison of taxonomic variability both within and between strata.

F. ALPHA, BETA, AND GAMMA DIVERSITIES

Whittaker (1972) has recently published an impressive review of different aspects of diversity, and provides an excellent synthesis of the ways in which diversity may have evolved. We feel however that we must take issue with him on several points. The first is semantic. Under the heading of species diversity and specifically under alpha diversity, he includes two concepts (p. 221) "(a) diversity proper, or richness of the community in numbers of species, and (b) the character of the importance/value relations. . . .." As indicated earlier we agree with Hurlbert (1971) that at the level of the first concept it is meaningful to distinguish between species richness and species density. We also believe the distinction made originally by Whittaker (1965) between "species diversity" and "dominance diversity" is worth maintaining.

The second objection is more fundamental. Alpha diversity is "community, within-habitat" diversity. It can only be a rigorous concept when either communities or habitats can be delineated in some way which finds general acceptance. The difficulties of setting boundaries to a habitat by human vision or instrumental aid are considerable, and as stressed earlier there is an ever-present risk that the boundaries of the species will not fit with where we have decided they should be. This is the whole difficulty about the biotopic concept. Much of our thesis boils down to a previous statement that only by a site classification based on species as attributes can we divide the biota into "communities" and the environment into site groups. There is no agreed level where we can stop our classification of sites, and we are quite ignorant of where in general terms a site classification and a "habitat" classification coincide.

This uncertainty becomes even more important when we distinguish beta diversity, which is the between-habitat diversity, from alpha diversity. Confusion is compounded over gamma diversity which Whittaker (1972, p. 231) says ". . . is consequent on the alpha diversity of the individual communities and the range of differentiation or beta diversity among them."

These criticisms apart, Whittaker should be consulted for his scholarly exposition and detailed documentation of relationships between diversity and environmental favorableness, instability, and climax development.

G. HABITAT WIDTH AND HABITAT OVERLAP

It can be shown (see Appendix) in a site/species situation that there are three ways in which total diversity can be partitioned (a) into diversity of (species-ignoring sites) + (sites-using species); (b) into (sites-ignoring species) + (species-using sites); (c) into (species-using sites) + (sites-using species) + interaction.

Pielou (1972) has shown that certain components of the above have ecological meaning and utility; thus, the weighted mean of the site diversity using all species is a measure of average habitat width, and similarly the weighted mean of the species diversity using all sites is a measure of average habitat overlap. Bradbury and Goeden (1975) have effectively applied these measures, using Brillouin diversity measures, to the study of resident fishes on a coral reef slope.

One suspects that variance measures could equally well be used for such partitioning. Even more simply, it seems that average intersite dissimilarities or interspecies dissimilarities as obtained by use of one of many of the dissimilarity measures (previously detailed), could be employed.

13

Multivariate Analysis

A. INTRODUCTION

The classificatory procedures described earlier would all result in the production of one or more two-dimensional graphs (dendrograms, minimum spanning trees, etc.) from a given set of data. In order to achieve such simplicity there has been of necessity a considerable loss of information. Much of this loss would be avoided if the data could be looked at in space of several dimensions. As this occasions difficulty beyond three dimensions, considerable attention has been given to the problem of determining the best condensation of multidimensional relationships by projection on to a reduced number of suitable planes. It should be noted that all models for achieving this purpose are basically linear. These methods of multivariate analysis are applicable to four quite distinct situations.

The first of these is *ordination* which makes no assumptions about the existence or otherwise of groupings among the entities and operates either on the attribute scores (*Principal component analysis*) or the dissimilarity matrix (*Principal coordinate analysis*). With the first, the positions of the sites or taxa are plotted initially against convention Cartesian coordinates; with the second, the interentity dissimilarities are plotted initially as distances into a set of orthogonal axes chosen such that the entity scores on the new axes preserve the interentity distances in the original system of axes.

Allied to these two ordinations are two other multivariate techniques whose purpose is also to investigate relationships in multidimensional space. However, in contrast to ordination with each of these, the data are already in groups. With *canonical variate analysis* the entities have been clustered and the relationships of these clusters to each other in multidimensional space are investigated; with *canonical correlation analysis,* the attributes instead of the entities are already in two groups and the axes are rotated so as to maximize the correlations between the groups.

Factor analysis is a term used in varying senses including that of Principal Component Analysis. Here the two are clearly distinguished. Though formerly little used by taxonomists or ecologists, the concept will be considered briefly below because the procedure in future will almost certainly gain favor for the study of weakly structured data.

B. PRINCIPAL COMPONENT ANALYSIS

Consider a set of entities defined in terms of a series of n axes, each representing a separate attribute.

It is the purpose of principal component analysis to choose axes within the multidimensional space in such a manner that the projections of the entities onto the axes will "best" display their relationships. The concept of "best" depends on the outlook of the viewer but it usually means in such a way that the entities are more widely separated one from another in terms of the new than in terms of the original axes.

By way of illustration consider the solar system of which the earth is a member. Here all planets whether viewed from each other or the sun move across the heavens in the plane of the ecliptic, passing and repassing one another in a most confusing manner, as witness the inability of astronomers for so long to interpret the structure of the system. If, however, the solar system is viewed from above or below, the relative arrangement of the planets and sun are readily appreciated. In other words, the position of the viewer relative to the system is important for an understanding of its structure.

The basic principles of principal component analysis as developed for two dimensions will now be considered in general terms. More detailed and mathematically vigorous treatments will be found in a number of texts including: Seal (1964), Blackith and Reyment (1971), and Morrison (1967). In Fig. 13.1 eleven entities (taxa or sites) have been plotted with respect to two attributes (x, y). From the figure it is clear that the two variables are correlated and that their "spread" along either axis will depend on the scale of measurement employed.

The word "scale" is here open to two interpretations. In the first instance, scale might be taken to mean the number of units of measurements in the axis. Thus the measurements in both attributes may have been in centimeters but plotted on different scales in that for the ordinate the intervals representing 1 cm were only one-half the length of the intervals representing

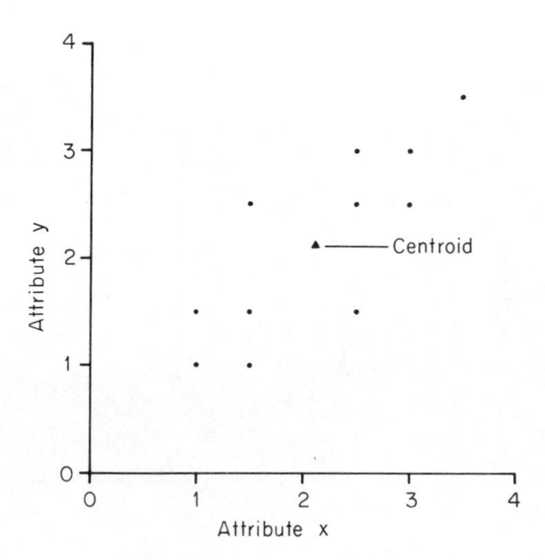

Fig. 13.1. The relationships between 11 entities compared with respect to two attributes.

1 cm of the abscissa. With the second usage of the term "scale," the original measurements may have been transformed before plotting and furthermore the transformation applied to the data for each set of attributes may have been different.

It is important that these two types of scaling be distinguished because of the differences which exist between regression and principal component analyses. Let it be assumed, with respect to Fig. 13.1, that it is desired to estimate the likely value of attribute y for a given individual for which the magnitude of attribute x is known. In this case a regression line is calculated; its direction being such that the score of squares of the vertical lines from the known pairs of coordinates to the lines is a minimum as shown in Fig. 13.2.

The most likely value of attribute y, given attribute x can now be estimated. Since with this prediction any values of attribute y are involved, the scale of attribute x is immaterial except in that it will determine the slope of the line. Thus for attribute y it does not matter whether the scale is compressed or expanded the same predicted value will be obtained. With this system attribute x is known as the independent and attribute y as the dependent variable.

Had it been desired that attribute x be predicted given attribute y, the "line of best fit" through the data would have been the line such that the sum of horizontal distances from the points of coordinates to the line was a

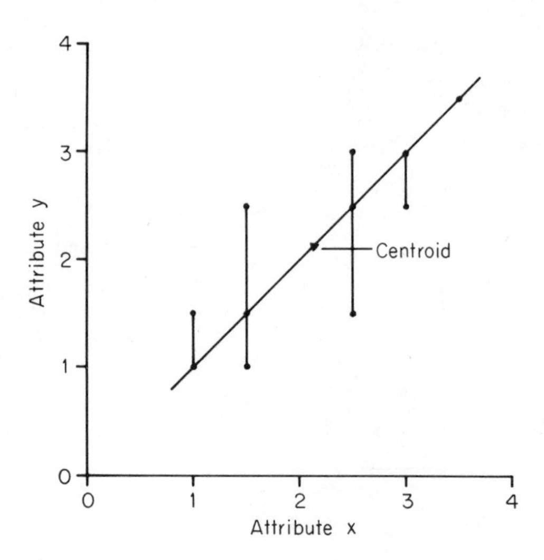

Fig. 13.2. The principle of fitting a regression line superimposed on the data of Fig. 13.1.

minimum. There are thus two regression lines possible, depending on which attribute is to be predicted. Both (regression) lines are given in Fig. 13.3.

In general it is assumed the independent attribute has fewer sources of error and hence less variability than the dependent, as, for instance, when weights are predicted given age or yields following the application of

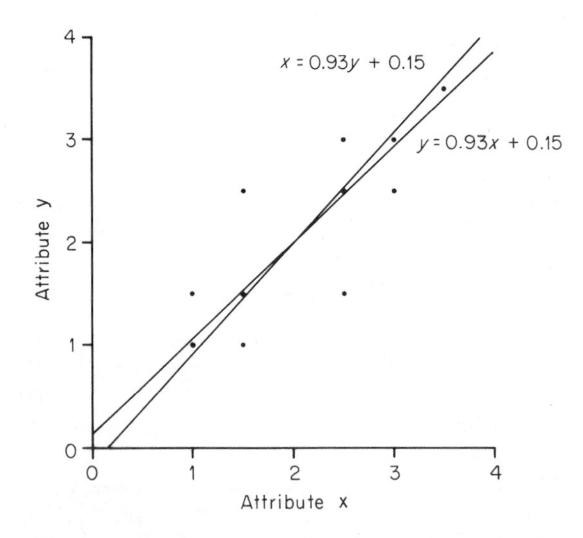

Fig. 13.3. The two possible regression lines appropriate to the data of Fig. 13.1.

stated amounts of fertilizer. With most taxonomic and ecological data derived from sampling populations of species, or areas sampled in terms of sites, none of the attributes may be taken as less "dependent" than the remainder. Hence, neither regression line is particularly appropriate. Instead it would seem the "line of best fit" is the one for which the sum of perpendiculars to the line from the pairs of coordinates is a minimum. One such perpendicular is shown in Fig. 13.4 for a line already fitted to a set of data.

From Fig. 13.4, it can be seen that the perpendicular distance from a point A (with coordinates x, y) to the line can be calculated in either of two ways. Thus $AD = AB \sin \theta$ or $AC \cos \theta$. As it is unreasonable for AD to have more than one set of dimensions it is clear both attributes should be of the same kind and measured in the same units. For instance it would cause mental confusion if attribute x was in grams and attribute y in centimeters, for then the definition of AD would be a mixture of both length and weight, since the trigonmetrical terms involve measurements parallel to both axes.

To overcome these objections all data should first be converted to dimensionless values. This objective is achieved by standardization to zero mean with units of standard deviation each side of it. An important consequence of such standardization is that each entity is now referable to a set of axes which pass through a point corresponding to the mean

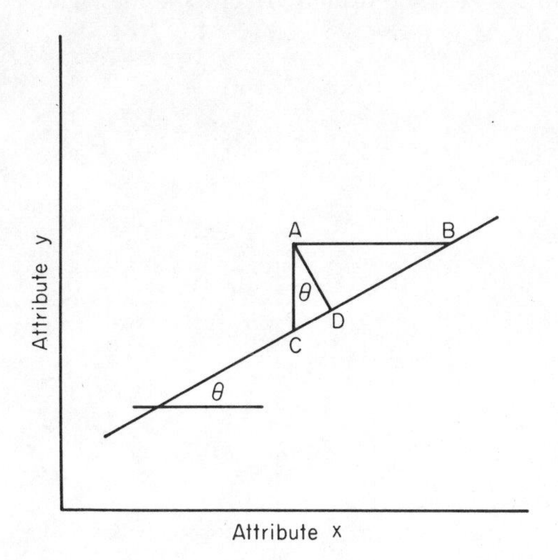

Fig. 13.4. An illustration of the principle of a line such that the sum of perpendiculars to the line is a minimum; this is only possible when the measurements along the two axes possess the same units.

value for each attribute. In terms of Fig. 13.1 this means the new axes would pass through the centroid of the entities, as plotted, and that each entity would have a set of new coordinates defined in terms of standard deviation units. This situation using the data from Fig. 13.1 is shown in Figure 13.5

If it is intended to display these data in such a manner as to achieve the maximum separation of entities in one dimension it is necessary for them to be projected onto a line and the problem to be solved is the position of a line to achieve this objective. The data could be projected onto either the ordinate or abscissa but neither of these would seem intuitively to be the desired line. Each of these axes represents a single variable and to accept either is tantamount to assuming the other contributes no additional information about the entity.

The appropriate line must lie between the two axes and as is readily proven it is the line resulting from minimizing the perpendiculars as discussed above. It follows that if the perpendicular distances from all coordinate pairs to the line are a minimum, the spread of these coordinate pairs as projected onto the line will be maximum. In support of the contention it can be confirmed from Fig. 13.6 that whereas the projections of the entities onto either the ordinate or the abscissa are spread over 3 scale units the projections onto the line of best fit as here defined extend over a spread of approximately 4.2 units.

The line along which the projections of the entities have a maximum spread is known as the first principal component, eigen vector, or latent

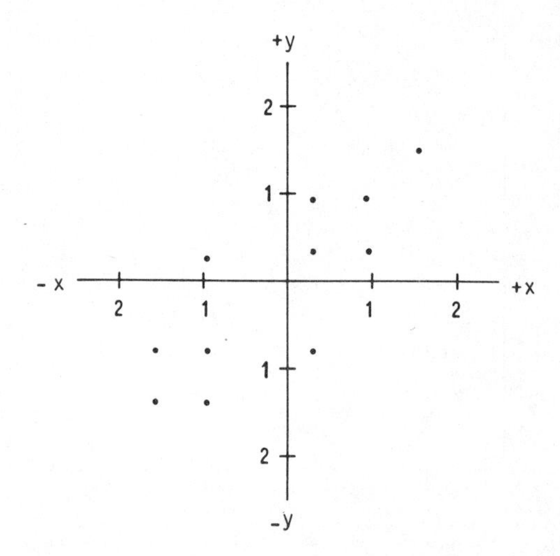

Fig. 13.5. The relationships between the entities of Fig. 13.1, plotted in standard deviation units against axes transferred to their centroid.

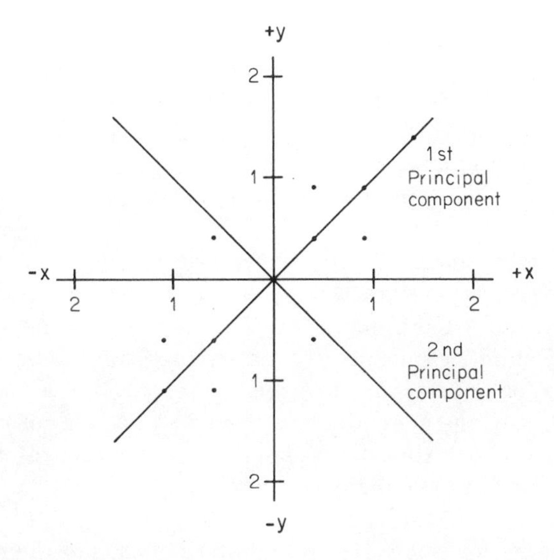

Fig. 13.6. Same as for Fig. 13.1 with the first and second principal component axes added to the diagram.

vector. The axis at right angles to this is known as the second principle component, eigen vector, or latent vector. It will be noted the scores of the entities as measured in terms of these two axes are quite independent of one another, and so the axes are described as orthogonal.

The concept of the spread of coordinates along the principle components may be made quantitative by calculating the variances of the coordinates with respect to each principle component. These variances are known as eigen-values or latent roots. The principal components can thus be distinguished in terms of the proportion of the total variance associated with each, and are generally subject to the further restriction that the first principal component must pass through the origin. With data that have been standardized and centered as above, this will automatically be so.

The coordinates or principal component scores for each entity expressed in terms of the principal components as axes are readily calculated from the original data. Three steps are involved. First, the original data are standardized to zero mean and unit variance. Second, the directions of the principal components are determined. Third, new scores, the principle component scores, are calculated from the original coordinates by referring the entities to the principal components instead of the original axes. In terms of analytical geometry this is the equivalent of rotating the axes.

Had the set of entities been defined with respect to three variables instead of two there would have been three principal components. The entities may then be projected onto any one of the three principal components or any of the three planes resulting from the axes being taken in pairs. As a rule not all of these projections are of equal interest.

In general the projection of most interest is that onto the plane defined by the pair of principal components which between them are associated with the two largest variances. These are the first and second principal components. The projection onto this plane will generally show the entities more discretely separated from one another than in any other projection.

Given a set of entities defined by n attributes, most computing centers have facilities for undertaking Principal Component Analyses. As stated above, the merit of the method is that it enables the relationships between a set of entities to be expressed in a space of fewer dimensions than that into which the entities are originally plotted.

While discrete or isolated groups may emerge when a set of entities is so projected, care must be taken in the interpretation of the results. Since what is seen is always a projection, the possibility of this involving several groups along the same line of sight must be clarified by reference to the coordinates for entities along other components.

The manner in which projections in various directions will sometimes expose and at other times obscure the existence of groups is shown in Fig. 13.7. Here there are three populations scored with respect to two continuous variables. In Fig. 13.7a the members of the populations are plotted against the original axes. If only one variable were available for these individuals it is clear that the existence of three separate populations would not be suspected. In Fig. 13.7b the original axes have been retained for reference and a new set, the principal components, inserted with their origin at the centroid of the system. From this figure it can be seen that with respect to the first principal component these individuals appear to belong to a single population, for their projection onto this axis would produce a continuous frequency distribution. However, if they are projected onto the second principal component it is clear that the total population comprises three distinct groups of individuals, which would project onto different parts of the axis.

While it is true that one of the objectives of principal component analysis is to express the relationships between entities in fewer dimensions than those available, this example demonstrates that in so doing information may be sacrificed and affinities may be obscured.

The account of principal component analysis assumes that the maximum separation of entities as projected onto certain axes or planes was the

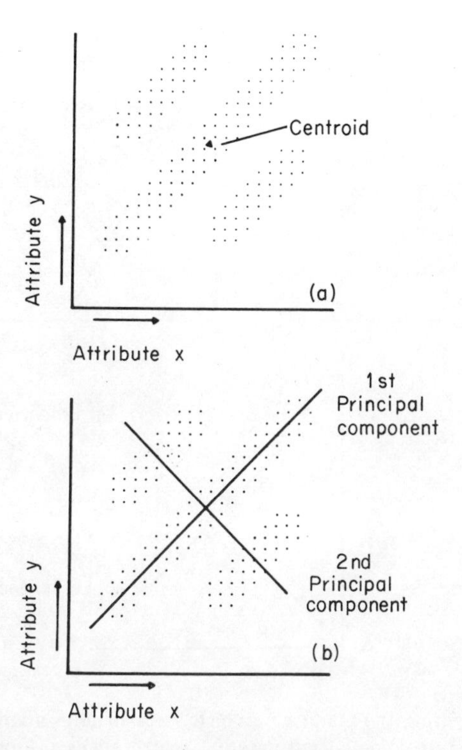

Fig. 13.7. The relationships between three populations with respect to two attributes, illustrating the manner in which projections onto different principal component axes may either reveal or conceal the existence of populations; (a) with reference to the original axes only; (b) with the principal component axes added.

objective of the operator and that furthermore one was restricted to working within the hyperspace containing the entities.

It would be possible to operate on other criteria such as maintaining the original origin and transforming or otherwise the original data (see Fig. 13.8). Such approaches as these may be of value to ecologists in particular and a wide variety of cases has been explored by Noy-Meir (1970, 1973).

A system for reducing data somewhat similar to that described above but computationally much simpler is due to Bray and Curtis (1957). They extract from the data the axis joining the most dissimilar pair of sites and then relate all other sites to these by projection onto that axis. It is generally agreed that principal component analysis is preferable but the Bray and Curtis approach may be necessary when the number of entities to be classified is large. (The advent of faster computers with larger storage facilities will shortly remove this restriction for most ecologists.)

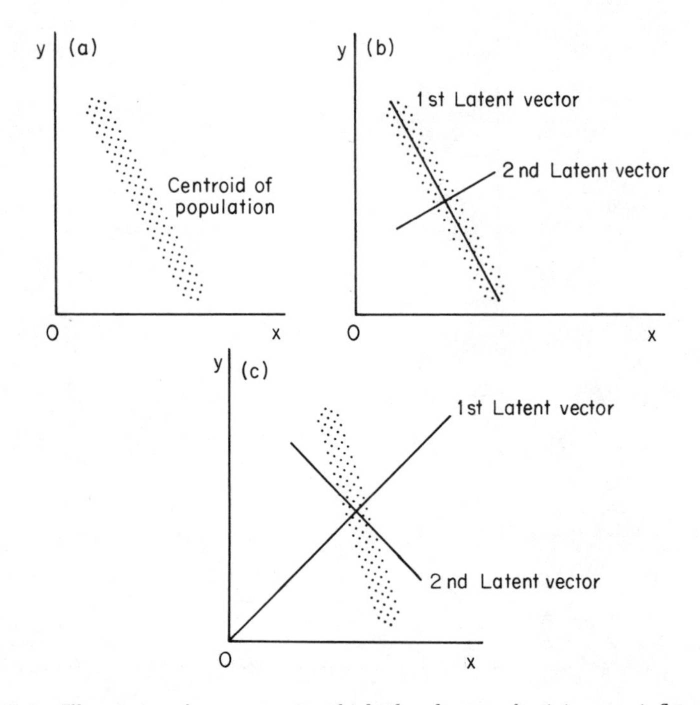

Fig. 13.8. Illustrating the manner in which the change of origin may influence the directions of the principal component axes; (a) scatter diagrams for population plotted against the original axes; (b) the same with principal component axes added and passing through the centroid of the population; (c) same as for (b) but with the principal component axes passing through the original origin and the centroid. (Note latent vector equals principal component.)

For a further discussion of this method, see Austin and Orloci (1966). Principal component analysis has an important place in taxonomic as well as ecological studies and has recently been applied to the study of intra-specific variation in *Acacia harpophylla* (Coaldrake, 1971); *Eucalyptus camaldulensis* (Burley *et al.*, 1971); and *E. maculata* (Andrew, 1970).

C. FACTOR ANALYSIS

As indicated above, principal component analysis begins with a set of observed entities scored for a number of attributes, and the principal components are sought in the expectation that relatively few of them will account for most of the variation in the population under study. Should this expectation prove to be so, the remaining principal components are neglected and "meanings" are often sought for those that are retained. These "meanings" will be discussed later in this section.

With factor analysis the reduction in the space required to account for most of the variation in the population is achieved by first regarding each attribute score as composed of two portions and then attempting to reduce the space necessary to express one of these adequately.

Let it be assumed that for a set of entities, attribute A is completely correlated with attribute B. A knowledge of the magnitude of either is now sufficient to predict accurately the other. Were A completely uncorrelated with B a knowledge of either attribute is of no value for predicting the other. In most instances, the attributes will be more or less correlated and so a knowledge of the magnitude of one will enable predictions to be made about the magnitude of the other with a given probability of being correct.

Thus, if a man weighs 100 kg it is highly probable he is taller than average, has large feet, a wider arm-span, etc. That is, each of these attributes reflects in some measure overall size. And so, for a given height H, part of the attribute is a reflection of overall size and part is not. In algebraic form this may be stated as: $H = (a + b)$ where a is a factor reflecting overall size and b is a factor independent of absolute size. (Here overall size has been measured as weight.)

If the original data have been reduced to standard form (zero mean, unit variance) the variance of H would be

$$\operatorname{var} H = 1 = \frac{1}{n-1}\left\{\sum_1^n (a+b)^2 - \left[\sum_1^n \frac{(a+b)}{n}\right]^2\right\}$$

$$= \frac{1}{n-1}\left\{\left[\sum_1^n a^2 - \frac{\left(\sum_1^n a\right)^2}{n}\right] + \left[\sum_1^n b^2 - \frac{\left(\sum_1^n b\right)^2}{n}\right]\right.$$

$$\left. + 2\left[\sum_1^n ab - \frac{\sum_1^n a \sum_1^n b}{n}\right]\right\}$$

where n is the number of individuals measured. If it is now assumed that a and b are uncorrelated, or they are defined so to be, the cross-products term in the above expression disappears (that portion within the last internal brackets) and the variance of H is seen to be composed of two portions. One of these is the common factor variance due to a, the other is the unique variance due to b.

The partitioning of all attributes into two factors as has been done for H, doubles the dimensions of the space to be considered, which might appear to run counter to the objective of reducing the space in which the original data may be represented. However, this apparent contradiction does not result because of the manner in which the partition has been undertaken. The common factor variance and unique variance components of the total variance for any given attribute are defined so as to be independent, which enables the rank of the joint dispersion matrix (variance–covariance matrix, usually standardized) to be reduced by suitable choice of common factor variances. The manner in which this may be achieved, though not the mechanism, will now be considered with reference to the two dispersion matrices shown in Table 13.1

In matrix I the variances and covariances between a set of 3 entities are given in standardized form. Because the data have been standardized the variances (diagonal entities) are all unity. The covariances (off diagonals)

TABLE 13.1

MATRICES I AND II[a]

Matrix I

	H^1	H^2	H^2
H^1	1.00	x	y
H^2	x	1.00	z
H^3	y	z	1.00

Matrix II

	a^1	a^2	a^3	b^1	b^2	b^3
a^1	α^1	x	y	0	0	0
a^2	x	α^2	z	0	0	0
a^3	y	z	α^3	0	0	0
b^1	0	0	0	β^1	0	0
b^2	0	0	0	0	β^2	0
b^3	0	0	0	0	0	β^3

[a] Where $H^1 = a^1 + b^1$; $H^2 = a^2 + b^2$; $H^3 = a^3 + b^3$ and α^1, α^2, α^3, β^1, β^2, and β^3 are communalities.

are all less than unity since with standardized data these are correlation coefficients. In matrix II the common factor variances, now known as communalities, are set in the diagonal cells of the upper left quarter and the unique variances in the diagonal cells of the lower right quarter of the enlarged matrix. Since the common factors and the unique factors are uncorrelated all cells of the upper right-hand and lower left-hand quarters contain zero values. Likewise, in the lower right-hand quarter all the correlation coefficients are zero since the unique factors must be uncorrelated. Finally, the common factor correlations may be shown to be identical with the corresponding correlations in matrix I.

It is the objective of factor analysis to find values of the communalities such that the correlation coefficients remaining in the matrix II are all reduced to zero. The rank of the matrix is thereby reduced and so automatically is the space in which the data may be represented. Except for a few matrices of given rank (e.g., 3×3 and $8 \times 8 \ldots$) for which the communalities may be calculated directly from the observed correlation coefficients, the communalities must be calculated indirectly through a series of iterations. The procedures for determining communalities are set out in texts on factor analysis (e.g., Harman, 1968).

Having determined the magnitude of the communalities the factors are obtained as their square roots. The proportion of the variance for which they account is then known. If the common factors account for only a small proportion of the total variance it may be decided to procede no further with the analysis. However, if the communalities of at least some of the factors are large the dispersion matrix may be used as the basis of a principal component analysis. Of the axes estimated those associated with unique factors may be neglected, and so the space in which the data are to be interpreted is reduced back to that corresponding to the original number of variables. It is then further reduced because in calculating the communalities the space required for expressing the common factors is automatically reduced. For further details, see Harman (1968).

Although factor analysis has been widely used by psychologists and educationalists it has been little used by either ecologists or taxonomists. Possibly this is because unlike the above two groups they had no basic theoretical model demanding its use. The method has been much criticized by some, largely on the basis of the interpretation of the results, and by others the method has been regarded as useless (Blackith and Reyment, 1971). Nonetheless there are occasions when factor analysis would seem preferable to principal component analysis, especially in ecological studies. This may be the case when the data include a large proportion of attributes only weakly intercorrelated, such as the near-random occurrence of certain species of low frequency. With standardized data each species contributes equally to the total variance of the population and so the proportion of the total variance extracted by the higher order principal components will of necessity be small. Indeed the proportion of the variance extracted by these principal components will be inversely proportional to the number of randomly distributed species.

In contrast to factor analysis, weakly correlated species pairs play only a minor role in determining the magnitudes of the communalities which are thereby relatively uninfluenced by them. It is likely that in the future factor analysis will play an increasingly important role in ecological studies.

Factor analysis has been applied to taxonomic problems, particularly by palaeontologists who have no access to living material and so must infer development from available specimens (Hemmings and Rostron, 1972). Additional examples are quoted by Blackith and Reyment (1971).

D. PRINCIPAL COORDINATE ANALYSIS

Unlike the situation discussed above, here there are no attributes to define the positions of the entities in multidimensional space. Such a situation is commonplace in classificatory studies, for it is common to generate matrices of interentity similarities or dissimilarities before starting a clustering strategy. Such a matrix can be conceived as representing a multidimensional space in which the entities are embedded in such a way that all are separated from one another by distances corresponding to their dissimilarities.

For a discussion on how to extract the principal components from a dissimilarity matrix and hence the coordinates along these components the reader is referred to Gower (1967, 1969).

The ability to ordinate a set of entities given only their dissimilarities is of particular value in many ecological studies. As shown above there are some circumstances in which certain dissimilarity measures are preferable to others. For example, one may wish to emphasize dominance and, hence, would use the Bray-Curtis measure or one may be concerned more with relative properties and so would wish to use the Canberra metric, both of which measures would generate a dissimilarity matrix of sites.

Principal coordinate analysis, therefore, gives one the opportunity to stress those features of the data thought to be of particular ecological importance. In contrast, principal component analysis is restricted always to the original data as axes and though the same stress as described above might be obtained by suitable scaling of the data prior to undertaking the principal component analysis, the form of the scaling to be employed is not generally clear. Furthermore, with most principal component analyses, the data are standardized and centered which means that all attributes become equally important in contributing to the total variance.

E. CANONICAL VARIATE ANALYSES

Unlike the multivariate analyses described above, canonical variate or multiple discrimination analysis begins with sets of sites or taxa already grouped and investigates the relationships between these groups. Since one begins with already defined clusters, canonical variate analysis is not properly a classificatory strategy and so has not been widely used by plant or animal taxonomists. It has, however, been extensively employed by physical anthropologists (Rao, 1952) and for specialized studies such as interspecific hybridization in *Eucalyptus* (Clifford and Binet, 1954) and incipient speciation in the long-tailed field mouse (Delany and Healy, 1964). It has also been used for displaying relationships amongst a sample of green algae (Ducker *et al.*, 1965) and between five taxa of gazelles (Rostron, 1972).

The objective of canonical variate analysis is to determine functions whose application to the original data (usually transformed and standardized) maximizes the observed differences between the groups. To this end each entity is assigned a total score secured by summation of each of the attribute scores multiplied by an appropriate weight. The weightings for each attribute usually differ in magnitude and this determination demands the prior calculation of the correlation matrix. The weights are so chosen as to ensure the variances of the total scores for entities is a minimum within groups. Hence the variances between groups must be a maximum.

In their study of hybridization in *Eucalyptus*, Clifford and Binet (1954) showed that whereas with respect to fruit length (mm), fruit weight (mg), and peduncle length (mm) the two species involved were largely similar; they were well differentiated in terms of the following function: $y = 16.12 \times \log_{10}$ (fruit length in mm) $+ 0.97 \times$ (fruit weight in mgm)$^{1/3}$ $+ 0.58 \times$ (peduncle length in mm)$^{1/2}$. Here y is the score for each individual in terms of its original measurements which have been transformed prior to the analysis.

F. CANONICAL CORRELATION ANALYSIS

Whereas with canonical variate analysis it is required to investigate the relationships between several groups given information on the attributes of the entities comprising them, with canonical variate analysis the interest is centered on the relationships between a series of entities given information about them in terms of two quite separate sets of attributes. For example, a pasture may have been sampled in terms of both botanical composition and the properties of the soil at a series of sites.

The extent of the agreement between the classification of sites based on each set of data separately would be a measure of the value of one set of data for predicting the other. Here the interest centers on the predictive value of whole sets of data and not on the ability of one attribute to predict another.

The capacity of an individual attribute in one set to predict the performance of an attribute in the other, though often of interest, is of little value here, for in most instances the number of correlations possible between the attributes of two sets is generally too large for them to be interpretable. Thus if a pasture possessed 50 species and at each site 10 soil properties were investigated it would be necessary to calculate 500 correlation coefficients, of which at least several might be spuriously large due to sampling errors alone. Hence, it is more useful to operate on the total data as a single set. This may be achieved as follows.

Imagine each site to be specified in terms of its species composition. If there are a total of m species, each site may be specified as a point in m-dimensional space, each axis representing a separate species. But each site is also defined in terms of n soil attributes and so may be specified as a point in n-dimensional space, each axis being a separate soil attribute. If the agreement between soils and vegetation is close the relative positions of the sites to one another in the two hyperspaces will be the same. The making of such comparisons demands that the sites be comparable on all axes and so the original data must first be centered and standardized. The estimation of the correlation between the two sets of data is then undertaken as follows.

Since the data must be in standardized form (zero mean and unit variance) all axes can be regarded as passing through a common origin. The axes of each set are now rotated until the original attribute scores are uncorrelated in terms of all axes within one set and with all save one, in the other set.

The canonical correlation axis resembles a principal component in that it is orthogonal to all other axes in its own set but suffers the further restriction that it is also orthogonal to all save one axis in the other set. The angle between the pair of correlated axes, one from each set, may be taken as a measure of the correlation between the sets. There are several possible pairs of such correlated axes and the correlations between them are known as canonical correlations. It is usual to study these canonical correlations in sequence beginning with the largest, and investigating only the first few. The cosine for the angle between the canonical correlation axes (those which satisfy the conditions outlined above) is equal to r, the product moment correlation coefficient.

Having determined the directions of the canonical correlation axes it is important to know what it is that they reflect concerning the original data. It will be remembered that the purpose of the analysis is to relate the two data sets applying to the same set of entities. The magnitude of the angles or correlation coefficients between the canonical correlation axis and the attribute axes in each set supply this information. Thus, the direction of the canonical correlation axis might be determined largely by the amount of only one species in the pasture or by several equally. That is, progression in one or the other direction from the origin along the canonical correlation axis might be paralleled by an increase or decrease of one or more species. This is similar to the situation met with in principal component analysis where as a rule, relatively few of the attributes determine the direction of the principal component.

Having determined for each of the two sets of data, the attributes within each that are most closely identified with the canonical correlation axes, these are also the attributes most closely correlated with each other.

Much of the earlier use of canonical correlation analysis was of the predictive type described above. However, the technique has a role to play in purely classificatory studies. Given the attributes most closely associated with the canonical correlation axes, these could be employed to provide a reduced space into which the sites may be plotted. This would be akin to an ordination procedure in that the canonical correlation analysis had been used to reduce the space in which the interrelating of the sites were to be investigated. There is, however, another potentially more useful extension of the technique.

Consider an ecological investigation in which there are a very large number of species and an equally large number of abiotic factors. Furthermore assume the data to be only weakly structured. In these circumstances the manipulation of the large matrices involved will be both expensive and unlikely to yield any attributes closely associated with the canonical correlation axes. It may be preferable to extract from each data set separately their first few principal components and to use the entity scores with respect to these principal components as attributes. This approach has been applied by Webb *et al.* (1971) to the investigation of the likely effective usage of rain forest areas in eastern Australia. The two data sets here were vegetational and environmental attributes. In a similar study aimed at the investigation of wide diversity in birds, Webb *et al.* (1973) employed as attributes principal coordinates rather than principal components by first ordinating their sites on the basis of intersite dissimilarities.

The employing of principal component or principal coordinate scores instead of original data may simplify the resultant canonical correlation analysis but it is equally likely to complicate the interpretation of the results. The meanings to be attached to principal component and principal coordinate axes depends often on personal interpretation unless they clearly reflect very few of the original variables. If on the one hand, they do so, these attributes might themselves be employed directly in the canonical correlation analysis. If on the other hand, the principal component reflects a large number of the original variables, then ordination prior to canonical correlation analysis would seem to be called for, even though at this stage further manipulation of the data would appear to make interpretation of the results more difficult. It is clear that further heuristic experience of ordination of attribute data prior to canonical correlation analysis is required.

For further theoretical background to this procedure, reference may be made to Williams and Lance (1968), Pielou (1969), and Cooley and Lohnes (1971).

G. INTERPRETING ORDINATIONS

The objective of ordination is to reduce the dimensions of the space in which the entities are presented. In so doing, certain lines or planes are chosen, and the original positions of the entities are projected on to these lines or planes. Such projections may lead to entities of groups which are quite different (e.g., above or below a given plane) projecting in such a manner as to suggest that they are the same (see Fig. 13.6). In this example by adding a further dimension, the entities will be separated. However, the addition of further dimensions in this fashion runs counter to the entire objective of ordination.

When data are complex it is often necessary to consider many principal components before all entities or groups of entities can be exposed; otherwise one group may be hidden behind another. For example, Lavarack

(1972) found it necessary to consider the first ten principal components in his taxonomic study of orchids, and Noy-Meir (1970, 1971) used about 20 axes in his study of Australian vegetation.

When ordinations are performed by computer, it is usual to have print-outs of the variance associated with the first few principal component axes. From inspection of these values, decisions can usually be made on the desirability of an ordination approach. If only a relatively small fraction comes out in the first three axes it seems probable that the original data have too complex a structure for the ordination approach to be of great immediate value.

One of the main advantages of ordination over classification is that relationships between entities which are obscured by two-dimensional presentation (for example, in dendrograms) may be revealed. The easiest method of presentation to reveal extra dimensions is to operate within the first three axes of an ordination and show the entities in three-dimensional space. The relevant models can then be converted for visualization and for publication in a two-dimensional form by means of perspective drawings. A taxonomic example occurs in Stephenson *et al.* (1968) and an ecological one in Stephenson and Williams (1971). In such drawings it is sometimes possible to introduce hints of relationships in both the fourth and fifth dimensions as was shown in Fig. 9.2.

It is now generally appreciated that there are many variants of classificatory methods which require that a choice be made, but it is less well known that there are many variants of ordination. Beals (1973) in his provocatively titled paper "Ordination: Mathematical Elegance and Ecological Naïveté" effectively restricts his comments to the shortcomings of principal component analysis, in comparison with the Bray and Curtis (1957) method. However, most of his comments are timely because principal component analysis is the only option commonly available. In brief, some of the main objections to principal component analysis relate to centering, standardizing, and transforming data.

Centering and standardizing are essential in ordinating sites using abiotic data, but if this technique is used when the data consist of records of species-in-sites it introduces problems. Consider two species, one ubiquitous and relatively abundant and the other occurring infrequently as single individuals. Converted to zero mean and units of standard deviation the values of the abundant species would become relatively homogenous and those of the infrequent species relatively heterogeneous. The latter would become much more important than the former in determining the principal components and clearly this offends ecological common sense. When there are numerous uncommon species as is usual in ecological survey, the overall picture would be dominated by conjoint absences. This, as stated earlier, is something to which ecologists in general find objection.

It is possible to ordinate data without their being centered. The disadvantages have been pointed out by Dagnelie (1960) and Orloci (1966), and the most important may well be that the first axis is virtually lost from the aspect of reducing the dimensionality of the data. It indicates only a "general factor." Noy-Meir (1970, 1971) has pointed out that if we center data it implies a main interest in deviations from average rather than average itself, and that this may or may not be so; Noy-Meir himself preferred noncentered data.

In the context of terrestrial vegetation, Beals (1973) states firstly that principal component analysis assumes that the stands represents a cluster of random points in a hyperellipsoidal space of normally varying density around the centroid. (In fact it is not necessary to have normal distribution of data for a principal component analysis directed toward classificatory ends.) Beals states second that it supposes a random sampling of the "cluster space." Dealing with the second point first, in previous remarks we have criticized random sampling in topographical space, but even if this were undertaken, as Beals points out, it is not necessarily related to the hyperellipsoidal space. With regard to the first point it is clear that if data are to be of "normally varying density" or have "multivariate-normal distribution" around the centroid, considerable attention should be given to transformation of data before it is centered and standardized. There appears to have been no consideration of this possibility in principal component analysis. Noy-Meir (1973) promises a further paper on "transformations" in relation to ordination, but it is believed this refers to standardizations as we have defined them. If species data were transformed prior to standardization, problems could be expected in view of Taylor's work (1961, 1971) or which indicates the likelihood that a different transformation would be required for each species.

The objections in regard to giving undue importance to rarer species can be met by using principal coordinate analysis. In this the data which are used in ordination are similar in form to those used in classification and are obtained as interentity dissimilarities, treated as distance. The data may have been transformed to reduce the importance of high values, or a dissimilarity measure which effectively downgrades the importance of rare species may be used, for example, the Bray-Curtis measure. Overall, it appears that this might remove some of the naivete from ordinations, to which Beals (1973) has objected.

There are, however, other consequences. One is that it now becomes essential to work from the centroid of the data and in Noy-Meir's terms we can only be interested in deviations from this. The second is that, if we work with a dissimilarity measure constrained between zero and unity (e.g., Canberra metric or Bray-Curtis), the space within the ordinations occur is more constrained than in the case of principal component analysis.

A general caution should be exercised when interpreting the results of an ordination. Having extracted the principal components from a dissimilarity or data matrix, it is tempting to regard them as reflecting underlying causes controlling the structure of the matrix itself and independent of the entities. Thus, in an ecological situation involving sites and species, the largest principal component might be reflecting some extrinsic abiotic factor in the environment whose influence is dominating the areal distribution of the species. In a taxonomic situation a similar principal component might be reflecting size or a common evolutionary history.

In an ecological context considerable attention has been given to this problem especially in an effort to determine the importance of the main abiotic variables controlling species distributions within communities or along continua. The extensive literature on this subject has been ably summarized by Noy-Meir (1970) and critical discussions on the topic have been offered by Swan (1970), Noy-Meir and Austin (1970), Austin and Noy-Meir (1971), and Beal (1973).

These writers stress that, as might be expected, there is not a linear relationship between the numbers of individuals of a species and its position along an environmental gradient. Furthermore, this situation obtains no matter what transformation of the data is undertaken since the general form of the "response curve" is bell shaped. From their studies of this problem Austin and Noy-Meir (1971, p. 771) conclude that "the axes or components of current ordination techniques can not in general be interpreted as the 'main vegetational gradients' in the commonly used sense of continuous sequences of sites and there is still less justification for interpreting them as the 'main environmental gradients'."

The interpretive aspects of ordination have always been open to dispute, some finding the techniques useful, others not finding them so. As with all numerical approaches to biology, care must be taken to avoid making assumptions about the material to be ordinated merely to satisfy the underlying assumptions required by the mathematics. It is very tempting to assume that because ordination reduces the space in which the data are to be presented, the system has been simplified. This is far from true for while the space has been reduced, the complexity of the axes has been increased. It is as if one has moved the problem from one of the understanding a complex space defined by many simple variables, to a simple space defined by several complex variables.

A decision usually avoided in ordination studies concerns tests of significance. It may be assumed that the data in hand are to be studied for their own sake and, hence, the analysis reveals relationships which might otherwise have been overlooked due to the large amounts of data to be interpreted. It might also have been assumed the data resulted from sampling an underlying parent population. This latter assumption may often lead to a welter of statistical difficulties and to the production of unconvincing statistical tests (Pielou, 1969).

Finally it may be observed that ordination and classifications are to a large extent complementary. A dendrogram is a two-dimensional representation of the data and may, due to group size dependency, give a distorted view of the data. A principal coordinate analysis likewise reduces the dimensions in which the data are presented and may also distort the true situation due to the inadequacies of projection.

In most studies it is not a matter of either classification or ordination but of both. This is because in the complex situations which demand numerical analysis it is desirable to view the data against a wide background of data reduction techniques. Then and then only, will it be possible, knowing the strengths and weaknesses of each technique, to be reasonably certain of correctly interpreting the situation being analyzed.

Appendix

A. INFORMATION THEORY MEASURES OF DIVERSITY

The process of classification can be regarded as creating order among the set of entities being classified. For example, animals may be divided into two large groups—vertebrates and invertebrates. Such a division produces two groups whose members resemble one another more than members of the other group. That is, the degree of order within the groups so recognized is greater than among their members before the groups were separated.

If the classificatory process is accomplished by agglomerating entities to generate a hierarchical clustering, at each step in the process the amount of disorder within groups increases. The similarity between classification and the partition of entropy in thermodynamics has led to the development of a series of information measures being devised for generating classifications in taxonomy and ecology.

The manner in which a measure of diversity may be derived for three simple models will now be considered.

Model 1

Suppose at a given site N individuals have been collected belonging to S species, and that of these n_j belong to jth species. Furthermore assume this site to be a random sample from an infinitely large population. The probability of selecting an individual of this species at random is n_j/N and the probability of selecting all n_j individuals is $(n_j/N)^{n_j}$. The probability of selecting the observed numbers of each of the other species may be similarly estimated. The products of these probabilities is their joint probability (P) and is the probability associated with the whole site. Hence,

$$P = \left(\frac{n_1}{N}\right)^{n_1} \times \left(\frac{n_2}{N}\right)^{n_2} \times \cdots \left(\frac{n_j}{N}\right)^{n_j} \times \cdots \times \left(\frac{n_S}{N}\right)^{n_S}$$

Converting to logarithms this expression becomes

$$\log P = n_1 \log n_1 - n_1 \log N + n_2 \log n_2 - n_2 \log N \cdots$$
$$+ n_j \log n_j - n_j \log n \cdots n_S \log n_S - n_S \log N$$
$$= \sum_1^S n \log n - \sum_1^S n \log N$$

But

$$\sum_1^S n = N$$

Therefore,

$$\log P = \sum_1^S n \log n - N \log N$$

Multiplying through by -1 the above expression for the diversity (H) becomes

$$H = N \log N - \sum_1^S n \log n \quad \text{where} \quad H = -\log P.$$

This formula was first proposed as a measure of diversity by Shannon (1948) after whom it is generally named. Its properties were further discussed by Shannon and Weaver (1949).

The properties of H may be readily appreciated by consideration of a few examples. When all individuals at a site belong to the one species $N = n$ and so $\sum_1^S n \log n = N \log N$ from which it follows H is zero. This relationship is as would be expected of a diversity measure because there is no diversity. If all species at the site are represented by only one individual, $\sum_1^S \log n$ will equal zero since log 1 is zero and so the site will exhibit its maximum diversity with $H = N \log N$. The diversities of a series of sites each possessing 100 individuals are given in Table A.1, calculations being based on logarithms to base e. From this it may be seen that the diversity is, as anticipated, a maximum when the species are present in equal proportions. The base employed is immaterial but all comparisons made must be the same.

Clearly the measure of diversity H varies from site to site according to the relative frequencies of the species, and while it has a lower limit of zero there is no absolute upper limit. As a consequence, comparisons are likely to be misleading unless N is the same for the sites to be compared.

Fortunately this problem is readily overcome by dividing the value of H obtained by the number of individuals at each site to obtain a diversity

TABLE A.1

THE DIVERSITY IN TERMS OF H TO THE BASE e IN EACH OF 7 SITES POSSESSING 100 INDIVIDUALS DISTRIBUTED AMONG 5 SPECIES

	Sites						
Species	1	2	3	4	5	6	7
1	80	70	20	60	50	100	96
2	5	10	20	15	25	0	1
3	5	5	20	15	15	0	1
4	5	10	20	15	5	0	1
5	5	5	20	15	5	0	1
H:	77.8	100.0	161.0	52.4	127.7	0.0	22.3

per individual instead of per site. Alternatively the proportions of each species at each site may be considered instead of their absolute numbers. This is equivalent to dividing each column of Table A.1 by the total number of species present at each site and calculating the resultant diversities in terms of p, the proportions of each species present.

In these circumstances $N \log N$ is eliminated, since it becomes zero, the formula for diversity becomes $H = \sum_1^S p \log p$. The differences between these two approaches to diversity per individual will now be discussed with reference to Table A.2. Here the two sites considered differ in that one has

TABLE A.2

TWO SITES DIFFERING IN NUMBERS OF INDIVIDUALS BUT CONSTANT WITH RESPECT TO THE SPECIES AND THEIR PROPERTIES[a,b]

	Site	
Species	1	2
1	6	12
2	3	6
3	1	2

[a] Logs to base e.

[b] The diversities of each of the two sites are as follows:

Site 1 $\quad 10 \log 10 - 6 \log 6 - 3 \log 3 - 1 \log 1 = 23.0 - 10.8 - 3.3 - 0.0 = 8.90$

Site 2 $\quad 20 \log 20 - 12 \log 12 - 6 \log 6 - 2 \log 2 = 59.9 - 29.8 - 10.8 - 1.4 = 17.90$

twice as many individuals as the other. Division of each of the site diversities by total number of individuals yield answers that are identical within the limits imposed by approximate calculations. Thus per site, the diversity per individual is 0.890 and for site 2 it is 0.895.

The diversities in terms of $p \log p$ are not calculated since it is obvious they will be identical.

Model 2

In the above discussion it was assumed that we are dealing with a sample only of the total population. If instead the sample is accepted as the total population it is more appropriate to use the Brillouin measure of information content or measure of diversity (Brillouin, 1962) which has been discussed in some detail by Pielou (1966, 1969).

The measure may be derived as follows: Given a population of N individuals comprising S species with n_1 of species 1, n_2 of species 2, . . . , n_j of species j, the probability of selecting at random a single individual of the jth species is n_j/N; and the probability that the next individual selected at random would also belong to the species is $n_j - 1/N - 1$, etc. Hence, the joint probability of selecting in succession all members of the jth species is

$$\frac{n_j}{(N)} \times \frac{(n_j - 1)}{(N - 1)} \times \cdots \frac{1}{(N - n_j + 1)} = \frac{n_j!}{(N)(N-1) \cdots (N - j + 1)}$$

and that of selecting the members of the other species may be similarly calculated. The product of these probabilities is the probability of selecting the observed population from the original sample of species. Hence,

$$P = \frac{n_1!}{(N - n_1 + 1)} \times \frac{n_2!}{(N - n_1 - n_2 + 1)} \times \cdots$$

$$= \frac{n_1! \times n_2! \cdots n_S!}{N!}$$

Taking logarithms and multiplying by -1 this expression becomes

$$-\log P = \log N! - \sum_{1}^{S} (\log n!)$$

which is the Brillouin measure of diversity.

In practice the difference between the Shannon and Brillouin measures of diversity are not important because $N \log N$ and $\log N!$ are monotonic.

With both the Shannon and Brillouin measures of diversity the numbers of each species at each site are considered and the structure of the measures is such that species not represented at a particular site contribute nothing. When sites are to be compared one with another joint absences may be of some interest.

Model 3

The derivation of a diversity measure based on the presence or absence of binary attributes only will now be considered in terms of a taxonomic example. Consider the following series of species scored for a single binary attribute.

Species	Attribute state
1	0
2	0
3	0
4	1
5	1

Here there is clearly more disorder or diversity in the set of species than there would be if each species had the same attribute score. It is required to find a measure of the diversity.

One such measure is the likelihood of obtaining a set of species with the attribute scores as observed. Such a likelihood may be derived as follows.

Let p be the probability of a species having the score zero and $1 - p$ be the probability of a species having the score unity. The likelihood (L) of obtaining the observed series of attribute scores for a_j species with the score zero and b_j species with the attribute score unity is

$$L = (p)^{a_j}(1 - p)^{b_j}$$

The term likelihood is applied to such an estimate of joint probability because in the absence of any presumption as to its value p must be estimated from the data. The most probable value will be the number of zero entries divided by the total number of entries. That is

$$p = \frac{a_j}{a_j + b_j} = \frac{a_j}{N} \qquad (\text{where } N = a_j + b_j)$$

and

$$1 - p = \frac{(a_j + b_j) - a_j}{a_j + b_j} = \frac{N - a_j}{N}$$

Hence, the likelihood of the observed series above is

$$L = \left(\frac{a_j}{N}\right)^{a_j} \left(\frac{N - a_j}{N}\right)^{N-a_j}$$

$$\begin{aligned}\log L &= a_j \log a_j - a_j \log N + (N - a_j) \log (N - a_j) \\ &\quad - (N - a_j) \log N \\ &= a_j \log a_j - a_j \log N + (N - a_j) \log (N - a_j) \\ &\quad - N \log N + a_j \log N\end{aligned}$$

After removal of the term $a_j \log N$ and rearranging

$$\log L = -N \log N + [a_j \log a_j + (N - a_j) \log (N - a_j)]$$

Multiplying by -1, the expression becomes

$$I = N \log N - [a_j \log a_j + (N - a_j) \log (N - a_j)]$$

where I replaces $-\log L$.

While with the Shannon diversity measure there is maximal value for a given N when all attributes contribute equally, the present measure is zero when the species are all alike with respect to the attribute score. Thus if a_j is equal to N or is itself zero, I becomes zero. The maximum value of I is attained when the attribute states are equally represented among the species. In these circumstances $a_j = N - a_j$ and $I = 2a_j(\log 2a_j - \log a_j)$.

When there are S attributes the individual attribute likelihoods are multiplied together to provide a measure of the likelihood of the set observed, and the value for diversity now becomes

$$I = SN \log N - \sum_{1}^{S} [a_j \log a_j + (N - a_j) \log (N - a_j)]$$

B. PARTITIONING OF DIVERSITY OF THE INFORMATION CONTENT OF A TWO-WAY TABLE

This is relevant to the incorporation of multistate data into dissimilarity measures, and also the use of diversity measures in ecology. We here refer not to its partitioning into quantitative and qualitative components (see Williams, 1972) but instead to its partitioning within a matrix containing, for example, meristic data on species in sites. This second type of partitioning can be effected by using either Shannon or Brillouin diversity. We will consider examples using the Shannon index, although realizing that other workers and particularly Pielou (1972) have used the Brillouin measure.

Consider the following matrix of species recordings in sites (matrix is within boxed area, derivatives outside).

		Sites (entities)			
		1	2	3	Σ
	1	10	11	11	32
Species	2	5	6	3	14
(attributes)	3	2	3	6	11
	Σ	17	20	20	57

We can first determine individual site diversities (Shannon) using the columns of data. For site 1 this is $17 \log 17 - (10 \log 10 + 5 \log 5 + 2 \log 2)$, which to base 10 equals 6.82, let this be symbolized by Cl (diversity of column 1). Similarly, C2 = C3 = 8.47.

We can also determine individual species diversities using the rows of values. For species 1 this is $32 \log 32 - (10 \log 10 + 2 \times 11 \log 11)$ which is 15.25, let this be symbolized by R1 (diversity of row 1). In a similar way R2 = 6.45 and R3 = 4.75.

We can also obtain the total diversity in the matrix, involving the nine numbers contained therein; this diversity, which we symbolize by T, is equal to 50.88.

Two other values can be calculated; the diversity of sites-ignoring species and of species-ignoring sites. The first is calculated from the sums of the columns (17, 20, and 20 in the above example) and is 27.13, the second is calculated from the sums of the rows (32, 14, and 11) and is 24.42. These five measures of diversity are readily related to one another in a manner similar to that of the analysis of variance.

The diversity among column totals represents the diversity between sites ignoring their species compositions. Within each column the diversity encountered is due to the different numbers of individuals of each species present. Cast in the same form as an analysis of variance this would be expressed as follows:

Total diversity = diversity between columns (sites-ignoring species) + diversity within columns (species within sites)
= 27.13 + 6.82 + 8.47 + 8.47 = 50.89

As calculated directly, the total diversity for the table was 50.88, and so the agreement between the two calculations can be accepted.

Now, the diversity among row totals represents the diversity between species ignoring sites while the variation within rows represents the diversity encountered in the numbers of the same species at different sites. Again cast as an analysis of variance this may be expressed as follows:

Total diversity = diversity between rows (species-ignoring sites) + diversity within rows (sites within species)
= 24.42 + 15.25 + 6.45 + 4.75 = 50.87

Thus the total diversity can be partitioned in two ways depending on whether one is interested primarily in the species diversities or site diversities. As Pielou (1972) has commented, the weighted mean diversity within rows, that is, between sites within species, is a measure of habitat width for individual species, while the weighted mean diversity within columns is a measure of average habitat overlap.

A further property of the data matrix not yet considered pertains to the information shown by the columns and rows. From an inspection of the table it is clear that a knowledge of the rows enables a better than average guess as to the column values for certain species. Thus, from a knowledge that a given species is species number 3 it would be adjudged as more likely to have come from site 3 than either of sites 1 or 2. Since the shared information is in both the rows and columns, it is in fact represented twice in the table. Hence the total may be expressed as:

Total diversity = diversity based on row totals + diversity based on columns totals − diversity shared by the rows and columns

The third term on the right hand side of the expression is usually known as the interaction and is a measure of interdependence exhibited between the sites and species. Thus, a high value for the interaction would indicate a given species to be common to particular sites while a low value would indicate the species numbers to be largely independent of the sites. Recasting of the above formula so the interaction term is expressed as a function of the remainder results in the following expression.

$$\text{Interaction} = \text{diversity between rows} + \text{diversity between columns} - \text{total diversity}$$

As applied to the above data table this becomes

$$\begin{aligned}\text{Interaction} &= 24.42 + 27.13 - 50.88 \\ &= 0.67\end{aligned}$$

The interaction may also be expressed as terms of the within row and column diversity. Thus in terms of the above data

$$\begin{aligned}\text{Interaction} &= \text{total diversity} - \text{within row diversity} \\ &\quad - \text{within column diversity} \\ &= 50.88 - 26.45 - 23.76 = 0.67\end{aligned}$$

The recognition and measurement of the interaction term permits the consideration of two further quantities, neither of which has been widely employed either by ecologists or taxonomists, though widely used by psychologists. These are the quantities equivocation and noise.

Assume it is desired to predict the row entries given the column entries. The accuracy of such a prediction will depend upon the relative amounts of diversity in the interaction and in the diversity between the columns. The difference between these two amounts of diversity is the *equivocation* and so may be defined as

$$\text{Equivocation} = \text{between row diversity} - \text{interaction diversity}$$

In terms of the above example,

$$\text{Equivocation} = 24.42 - 0.67 = 23.75$$

Noise is a property similar to equivocation but is the difference between the row diversity and the interaction diversity.

There is no consistent usage of the terms equivocation and noise with respect to rows and columns so if these were interchanged the noise in one table would be the equivocation in the other table.

The principles employed for partitioning diversity have already been discussed. Here the formulas appropriate to such partitions will be generalized with respect to a taxonomic example and a few of their properties considered. In passing it must be mentioned that in general taxonomists have used the term information content where ecologists have employed the term diversity.

For simplicity the formulas are developed below for a 3 $\times$ 3 table; they

TABLE A.3

THIRTY INDIVIDUALS CLASSIFIED BOTH ACCORDING TO FLOWER COLOR AND SPECIES GROUPING

	Species group		
Attribute state	1	2	3
Red	2	2	1
White	5	2	6
Blue	3	6	3

apply equally to all two-way tables. Consider Table A.3, or its more general form as Table A.4.

Certain components of diversity are given below (1)–(4)

TABLE A.4

A GENERALIZED 5 × 3 TABLE[a]

	Column			
Row	1	2	3	
1	a	b	c	R_1
2	d	e	f	R_2
3	g	h	i	R_3
	C_1	C_2	C_3	N

[a] Where $R_1 = a + b + c$; $R_2 = d + e + f$; and $R_3 = g + h + i$. $C_1 = a + d + g$; $C_2 = b + e + h$; and $C_3 = c + f + i$. and $N = a + b + c + d + e + f + g + h + i$.

1. DIVERSITY WITHIN ROWS

For the attribute state red the diversity between species groups is

$$5 \log 5 - 2 \log 2 - 2 \log 2 - 1 \log 1$$

or in the general case

$$R_1 \log R_1 - a \log a - b \log b - c \log c$$

In the general case, summed over the three attribute states this becomes

$$R_1 \log R_1 + R_2 \log R_2 + R_3 \log R_3 - a \log a \cdots - i \log i$$

2. Diversity between Rows

Here the existence of species groups is ignored and the diversity estimated on the basis of the total number of red, blue, and white flowered individuals. In the general case, the diversity between rows is

$$N \log N - R_1 \log R_1 - R_2 \log R_2 - R_3 \log R_3$$

3. Diversity within Columns

This is a reflection of the different numbers of red, white, and blue flowers belonging to each species group and for the first species group would be

$$10 \log 10 - 2 \log 2 - 5 \log 5 \log 5 - 3 \log 3$$

In the general case this is

$$C_1 \log C_1 - a \log a - d \log d - g \log g$$

Summed over all attributes this becomes

$$C_1 \log C_1 + C_2 \log C_2 + C_3 \log C_3 - a \log a \ldots i \log i$$

4. Diversity between Columns

Here the differences in color are ignored and the diversity is estimated on the basis of the total number of entities in each species group. That is the diversity between rows is

$$N \log N - C_1 \log C_1 - C_2 \log C_2 - C_3 \log C_3$$

The relationship between the partition of diversity and the analysis of variance is evident if, for either the rows or columns, the appropriate within and between diversities are summed. When this is done the total diversity is attained. Thus by summing the within and between row diversities the following expression is obtained

$$N \log N - R_1 \log R_1 - R_2 \log R_2 - R_3 \log R_3 + R_1 \log R_1 - a \log a$$
$$- b \log b - c \log c + R_2 \log R_2 - d \log d - e \log e - f \log f$$
$$+ R_3 \log R_3 - g \log g - h \log h - i \log i$$

which on simplification becomes the expression for the total diversity.

5. Interaction Diversity

When calculating the above diversities no attention has been given to a joint consideration of the rows and columns. If the entries in each were fully independent the diversity of the rows and columns would be quite independent. However, if some of the diversity in the rows is due to diversity in the columns and vice versa they must be regarded as interacting. The estimation of the interaction diversity can be undertaken in several ways of which the following is perhaps the simplest. Since some of the diversity in the rows is shared with the columns and some of the diversity in the columns is shared with the rows the diversity within each will be less than it might otherwise have been. Hence, the interaction diversity may be found as the difference between the total diversity and the sum of the within row and within column diversities.

In terms of a 3×3 table this partition of diversity is given in Table A.5. It should be noted that the interaction diversity belongs equally to the rows or columns, not in the sense of being shared in even amounts, but belonging totally to each. Some properties of the interaction diversity will now be discussed.

The interaction diversity may vary in magnitude from zero to an indefinitely large value. Two particular examples follow.

	Species group				Species group		
Attribute state	1	2	3	Attribute state	1	2	3
Red	10	10	10	Red	20	5	5
White	10	10	10	White	5	20	5
Blue	10	10	10	Blue	5	5	20

Where the number of individuals is the same in each cell as in the left-hand table the interaction is zero, where the numbers differ as in the right-hand

TABLE A.5

PARTITION OF INFORMATION CONTENT IN A 3×3 TABLE

Within rows	$= R_1 \log R_1 + R_2 \log R_2 + R_3 \log R_3 - a \log a - b \log b - c \log c \ldots.$
Within columns	$= C_1 \log C_1 + C_2 \log C_2 + C_3 \log C_3 - a \log a - b \log b - c \log c \ldots.$
Interaction	$= N \log N - R_1 \log R_1 - R_2 \log R_2 - R_3 \log R_3 - C_1 \log C_1 - C_2 \log C_2 - C_3 \log C_3 + a \log a + b \log b \ldots\ldots.$
Total	$= N \log N - a \log a - b \log b - c \log c \ldots\ldots$

table the interaction is considerable being 20.77 (to base e) as calculated from the appropriate formula of Table A.5

Furthermore, it is not essential for the numbers of entries in a table to be identical for the interaction to be zero. It is sufficient that the row or column entries bear a constant ratio one to another. Thus in the following table there is no interaction.

Attribute state	Species group 1	2	3
Red	1	2	4
White	2	4	8
Blue	1	2	4

This is a useful property of the interaction diversity which permits its use as a measure of gain in diversity or information when combining multistate attributes as will be discussed below. Meanwhile it is sufficient to note that as expected there is no increase in interaction diversity if differing numbers of entities in two groups are fused to form a single group, provided their members possess attribute states in the same proportions.

C. INFORMATION GAIN WITH MULTISTATE ATTRIBUTES

Here the manner in which the information gain resulting from fusing pairs of groups with respect to multistate (including continuous) attributes will be discussed. These are not the only ways in which the problem may be approached, but this is the manner in which multistate data have been incorporated into such information gain classificatory programs as MULTBET mentioned in Chapter 8 Section D,2.

1. Disordered Multistate Attributes

Let the disordered multistate attribute possess four states and let each of the species groups (a and b) to be fused possess six members each with the attribute states distributed among the entities of the groups as in the following table:

Species group	Attribute state			
	1	2	3	4
a	—	4	1	1
b	4	1	—	1

The information within the rows and columns of the table is of no interest for we require to know the gain in information resulting from fusing the two groups. This is clearly the transmitted information or interaction diversity in the table and is calculated according to the formula derived above and listed in Table A.5. The information gain on fusing in the above case is then 16.5 (to base e).

2. Ordered Multistate Attributes

Here the sequence of the states is of relevance indicating the relative magnitudes of the attribute states. Here again it is transmitted information that is of interest and the previous table will suffice to illustrate the treatment of ordered multistate attributes it being borne in mind that their status has changed.

The information gain resulting from fusing groups a and b is calculated by reducing the number of attribute states to two as follows. Attribute 1 is contrasted with attributes $2 + 3 + 4$ as in the following table:

Species group	Attribute states	
	1	$2 + 3 + 4$
a	—	6
b	4	2

The interaction is here 25.9.

The attribute states are then contrasted as follows:

Species group	Attribute state	
	1 + 2	3 + 4
a	4	2
b	5	1

where the information gain is 0.1.

Finally the attribute states are contrasted as in the following table where states 1, 2, and 3 are compared with state

Species group	Attribute state	
	1 + 2 + 3	4
a	5	1
b	5	1

Here the interaction is 0.0, which is as would be expected since the numbers of elements within columns are the same for both group *a* and group *b*.

The information gain on fusing is then taken as that associated with division of the original table which gives the greatest gain on fusing, had the attribute possessed only two states. In the above case the division is 1 – 2, 3, 4.

3. Continuous Attributes

These are treated as above by dividing the range into a fixed number of states and noting the number of individual elements in each. The number of states commonly recognized is 8 and as with ordered multistate treatment just described only those pairs of comparisons resulting from dividing the series from left to right are considered.

D. INFORMATION MEASURES AND INTERDEPENDENCE

It is usual for entities to be defined in terms of several attributes that are not physically related. Thus, flowering plants can be classified by apparently independent disordered multistates such as fruit type, life form, and so on. Nonetheless correlations may exist among the attributes in the sense that a knowledge of the state of one may be useful for predicting the state of another. Information measures are useful here in that any attributes may be compared provided they possess discrete states.

In their study of *Arceuthobium*, Hawkesworth *et al.* (1968) were interested in the degree to which attributes were interdependent and for this purpose used an index proposed by Estabrook (1967). His measure of interdependence I_{AB} between attributes A and B was

$$I_{AB} = \frac{\text{Information held exclusively by } (A) \text{ plus information held exclusively by } (B)}{\text{total information possessed by both } (A) \text{ and } (B)}$$

In terms of the symbolism employed herein this may be rewritten as:

$$I_{AB} = \frac{\text{Diversity in } (A) + \text{diversity in } (B) - 2 \text{ interaction } (AB)}{\text{total diversity possessed by } (A) \text{ and } (B)}$$

The index I_{AB} is constrained between 0 when the two attributes are completely interdependent and 1 when a knowledge of the state of one contributes nothing to a knowledge of the state of the other.

The measure applies to two-way tables with any number of rows and columns but for convenience its applicability will be demonstrated with respect to a 2×2 table. Consider the following 2×2 table where two binary attributes have been recorded for N entities:

		Attribute A		
		+	−	
Attribute B	+	a	b	$a + b = R_1$
	−	c	d	$c + d = R_1$
		$a + c$	$b + d$	$a + b + c + d = N$
		$a + c$	$b + d$	
		C_1	C_2	

Applying the results of the above table to estimate the interdependence of

A and B the following result is obtained:

$$\mathrm{I}_{AB} = \frac{\begin{array}{l} 2(N\log N - a\log a - b\log b - c\log c - d\log d) \\ \quad -(N\log N - R_1\log R_1 - R_2\log R_2) \\ \quad -(N\log N - C_1\log C_1 - C_2\log C_2)\end{array}}{N\log N - a\log a - b\log b - c\log c - d\log d}$$

$$= \frac{\begin{array}{l} R_1\log R_1 + R_2\log R_2 + C_1\log C_1 + C_2\log C_2 \\ \quad -2(a\log a + b\log b + c\log c + d\log d)\end{array}}{N\log N - a\log a - b\log b - c\log c - d\log d}$$

when $a = b = c = d$

$$I_{AB} = \frac{4 \times 2a\log 2a - 2(4a\log a)}{4a\log 4a - 4a\log a}$$

$$= \frac{8a\log 2 + 8a\log a - 8a\log a}{8a\log 2 + 4a\log a - 4a\log a} = 1$$

when $b = c = 0$ or with appropriate change of symbols $a = d = 0$

$$I_{AB} = \frac{2a\log a + 2d\log d - 2(a\log a + d\log d)}{(a + d)\log(a + d) - a\log a - d\log d} = 0$$

It should be noted that I_{AB} is an estimate of the extent to which attributes or entities are interdependent in the sense of being useful for predicting the states of one another in each of the entities. Complete interdependence results when either $b = c = 0$ or $a = d = 0$. In the former instance the two species are completely alike in that whenever either has an attribute in either the positive or negative state so does the other; in the latter case the two species are totally unalike in that whenever one has an attribute in the positive state the other possesses it in the negative state and vice versa.

As a measure of association the failure of this index to distinguish between positive and negative associations is unimportant in that a knowledge as to the state of a given attribute may be just as valuable a predicter of the presence as of the absence of the state of another attribute. However it does limit the circumstances under which I_{AB} may be employed as an index of similarity.

It has been reported by Estabrook (1967) that the similarity index (S.I.) defined below is a proper metric.

$$\text{S.I.} = \sqrt{1 - I_{AB}^2}$$

It will be noted that when $I_{AB} = 0$, S.I. = 1 and that when $I_{AB} = 1$, S.I. = 0.

Bibliography

Adams, C. C. (1909). The ecological succession of birds. *Mich. Geol. Surv., Annu. Rep.* pp. 121–154.

Ager, D. V. (1956). Geographical factors in the definition of fossil species. *In* "The Species Concept in Paleontology" (P. C. Sylvester-Bradley, ed.), pp. 105–109. Systematics Association, London.

Allee, W. C., Emerson, A. E., Park, O., Park, T., and Schmidt, K. P. (1949). "Principles of Animal Ecology." Saunders, Philadelphia, Pennsylvania.

Anderson, A. J. B. (1966). A review of some recent developments in numerical taxonomy. M.Sc. Thesis, University of Aberdeen, Scotland.

Andrew, I. A. (1970). Inter-provenance variation of *Euc. maculata* Hook. grown in Zambia. *Aust. Forest.* **34,** 192–202.

Anonymous. (1959). Final resolution. Symposium on the Classification of Brackish Waters, Venezia 8–14 Aprile, 1958. *Arch. Oceanogr. Limnol.* **11,** Suppl., 243–245.

Anonymous.(1973). IMER. Continuous plankton records: A plankton atlas of the North Atlantic and the North Sea. *Bull. Mar. Ecol.* **7,** 1–174; see also bibliography **7,** XI–XIX.

Austin, M. P., and Greig-Smith, P. (1968). The application of quantitative methods to vegetation survey. II. Some methodological problems of data from rainforest. *J. Ecol.* **56,** 827–844.

Austin, M. P., and Noy-Meir, I. (1971). The problem of nonlinearity in ordination: Experiments with two-gradient models. *J. Ecol.* **59,** 763–773.

Austin, M. P., and Orloci, L. (1966). Geometric models in ecology. II. An evaluation of some ordination techniques. *J. Ecol.* **54,** 217–227.

Barrett, G. W. (1968). The effects of an acute insecticide stress on a semi-enclosed grassland ecosystem. *Ecology* **49,** 1019–1035.

Bartlett, H. H. (1940). History of the generic concept in botany. *Bull. Torrey Bot. Club* **67,** 349–362.

Bayer, F. M. Voss, G. L., and Robins, C. R. (1970). Report on the marine fauna and benthic shelf-slope communities of the Isthmian region. *Battelle Mem. Inst., Ohio* pp. 1–94.

Beals, E. W. (1973). Ordination: Mathematical elegance and ecological naïveté. *J. Ecol.* **61,** 23–35.

Benzecri, J.-P. (1969). Statistical analysis as a tool to make patterns emerge from data. *In* "Methodologies of Pattern Recognition" (S. Watanabe, ed.), pp. 35–74. Academic Press, New York.

Blackith, R. E., and Reyment, R. A., eds. (1971). "Multivariate Morphometrics." Academic Press, New York.

Boesch, D. F. (1971). Distribution and structure of benthic communities in a gradient estuary. Ph.D. Thesis, Virginia Institute of Marine Science, Gloucester Point, Virginia.

Bor, N. L. (1960). "The Grasses of Burma, Ceylon, India and Pakistan," Pergamon, Oxford.

Boudouresque, C. F. (1970). Recherches sur les concepts de biocoenose et de continuum au niveau de peuplements benthiques sciaphiles. *Vie Milieu, Ser. B* **21,** 103–136.

Bowman, T. E. (1971). The distribution of calanoid copepods off the southeastern United States between Cape Hatteras and Southern Florida. *Smithson. Contrib. Zool.* **96,** 1–58.

Boyce, A. J. (1969). Mapping diversity: A comparative study of some numerical methods. *In* "Numerical Taxonomy" (A. J. Cole, ed.), pp. 1–31. Academic Press, New York.

Bradbury, R. H., and Goeden, G. B. (1975). The partitioning of the reef slope environment by resident fishes. *Proc. Int. Symp. Corals, Coral Reefs, 2nd* Vol. 1, Gr. Barrier Reef Committee, Brisbane (in press).

Braun-Blanquet, J. (1932). "Plant Sociology; the Study of Plant Communities" (translated and edited by G. D. Fuller and H. C. Conrad). McGraw–Hill, New York.

Braun-Blanquet, J. (1951). "Pflanzensoziologie," 2nd ed. Springer-Verlag, Berlin and New York.

Bray, J. R., and Curtis, J. T. (1957). An ordination of the upland forest communities of southern Wisconsin. *Ecol. Monogr.* **27,** 325–349.

Brillouin, L. (1962). "Science and Information Theory," 2nd ed. Academic Press, New York.

Brinckmann, R. (1929). Statistisch-biostrategraphische Untersuchungen an Mittlejurassischen Ammoniten über Artbeguft und Stammesentwicklung. *Abh. Ges. Wiss. Goettingen Math.-Phys. Kl.* [N. S.] **13,** 1–249.

Brockmann-Jerosch, H. (1907). "Die Pflanzengesellschaften der Schweizeralpen. I. Die Flora des Puschlav (Berzirk Bernina, Kanton Graubunden) und ihre Pflanzengesellschaften." Engelmann, Leipzig.

Burley, J., Woodward, P. J., and Hans, A. S. (1971). Variation in leaf characteristics among provinances of *Eucalyptus camaldulensis* Dehn. grown in Zambia. *Aust. J. Bot.* **19,** 237–249.

Burr, E. J. (1968). Cluster sorting with mixed character types. I. Standardization of character values. *Aust. Comput. J.* **1,** 97–99.

Burr, E. J. (1970). Cluster sorting with mixed character types. II. Fusion strategies. *Aust. Comput. J.* **2,** 98–103.

Burt, R. L., Edye, L. A., Williams, W. T., Grof, B., and Nicholson, C. H. L. (1971). Numerical analysis of variation patterns in the genus *Stylosanthes* as an aid to plant introduction and assessment. *Aust. J. Agr. Res.* **22,** 737–757.

Buzas, M. A., and Gibson, T. G. (1969). Species diversity: Benthic Foraminifera in the western North Atlantic. *Science* **163,** 72–75.

Cain, A. J. (1953). Geography, ecology and co-existence in relation to the biological definition of the species. *Evolution* **7,** 76–83.

Cain, A. J. (1954). "Animal Species and Their Evolution." Hutchinson, London.

Cain, A. J., and Harrison, G. A. (1958). An analysis of the taxonomist's judgement of affinity. *Proc. Zool. Soc. London* **131,** 85–98.

Campbell, B. M., and Stephenson, W. (1970). The sub-littoral Brachyura (Crustacea: Decapoda) of Moreton Bay. *Mem. Queensl. Mus.* **5,** 235–301.

Carriker, M. R. (1967). Ecology of estuarine benthic invertebrates: A perspective. *In* "Estuaries," Publ. No. 83, pp. 442–487. Amer. Ass. Advan. Sci., Washington, D.C.

Cassie, R. M. (1961). The correlation coefficient as an index of ecological affinities in plankton populations. *Mem. Ist. Ital. Idrobiol. de Marchi* **13,** 151–177.

Cassie, R. M., and Michael, A. D. (1968). Fauna and sediments of an intertidal mud flat: A multivariate analysis. *J. Exp. Mar. Biol. Ecol.* **2,** 1–23.

Cattell, B. (1952). "Factor Analysis." Harper, New York.

Chace, F. C. (1969). Unknown species in the sea. *Science* **163,** 1271.

Cheetham, A. H., and Hazel, J. E. (1969). Binary (presence/absence) similarity coefficients. *J. Paleontol.* **43,** 1130–1136.

Clayton, W. D. (1970). Studies in the Gramineae. xxi. *Coelorhachis* and *Rhytachne*: A study in numerical taxonomy. *Kew Bull.* **24,** 309–314.

Clayton, W. D. (1972). Studies in the Gramineae. xxvi. Numerical taxonomy of the Arundinelleae. *Kew Bull.* **26,** 111–123.

Clements, F. E. (1905). "Research Methods in Ecology." Univ. Publ. Co. Lincoln, Nebraska.

Clements, F. E., and Shelford, V. E. (1939). "Bioecology." Wiley, New York.

Clifford, H. T. (1965). The classification of the Poaceae: A statistical study. *Univ. Queensl. Pap., Dep. Bot.* **4,** 241.

Clifford, H. T., and Binet, F. E. (1954). A quantitative study of a presumed hybrid swarm between *Eucalyptus elaeophora* and *E. goniocalyx*. *Aust. J. Bot.* **2,** 325–336.

Clifford, H. T., and Goodall, D. W. (1967). A numerical contribution to the classification of the Poaceae. *Aust. J. Bot.* **15,** 499–519.

Clifford, H. T., Williams, W. T., and Lance, G. N. (1969). A further numerical contribution to the classification of the Poaceae. *Aust. J. Bot.* **17,** 119–131.

Coaldrake, J. E. (1971). Variation in some floral, seed growth characteristics of *Acacia harpophylla* (brigalow). *Aust. J. Bot.* **19,** 335–352.

Connor, D. J., and Clifford, H. T. (1972). The vegetation near Brown Lake, North Stradbroke Island. *Proc. Roy. Soc. Queensl.* **83,** 69–82.

Cooley, W. W., and Lohnes, P. R. (1971). "Multivariate Data Analysis." Wiley, New York.

Cormack, R. M. (1971). A review of classification. *J. Roy. Statist. Soc., Ser. A* **134,** 321–367.

Corner, E. J. H. (1952). "Wayside Trees of Malaya," 2nd ed. Govt. Printing Office, Singapore.

Correll, R. L. (1974). The application of numerical taxonomy to the family Cyperaceae. Ph.D. Thesis, James Cook University of North Queensland.

Cowles, H. C. (1899). The ecological relations of the vegetation of the sand dunes of Lake Michigan. *Bot. Gaz.* (*Chicago*) **27,** 95–117, 167–202, 218–308, and 361–391.

Cragg, J. B. (1953). Book review of L. R. Dice, "Natural Communities." *Bull. Inst. Biol.* **1,** 3.

Crisp, D. J., and Southward, A. J. (1958). The distribution of intertidal organisms along the coasts of the English Channel. *J. Mar. Biol. Ass. U.K.* **37,** 157–208.

Czekanowski, J. (1913). "Zarys Metod Statystycznck." E. Wendego, Warsaw; see also 'Coefficient of racial likeness' and 'Durchschnittliche Differenz.' *Anthropol. Anz.* **9,** 227–249 (1932).

Dagnelie, P. (1960). Contribution à l'étude des communautés végétales par l'analyse factorielle. *Bull. Serv. Carte Phytogeogr., Ser. B* **5,** 7–71 and 93–195.

Dagnelie, P. (1966). A propos des différentes méthods de classification numerique. *Rev. Statist. Appl.* **14,** 55–75.

Dahl, E., Gjems, O., and Kielland-Lund, J. (1967). On the vegetation types of Norwegian conifer forests in relation to the chemical properties of the humus layer. *Medd. Nor. Skogsforsoeks.* **23,** 505–531.

Dahl, F. (1908). Die Lycosiden oder Wolfspinnen Deutschlands und ihre stellung im Haushalte der Natur. *Nova Acta Leopold. Carol. Deut. Akad. Naturforsch.* **88,** 174–178.

Darlington, P. J. (1957). "Zoogeography: The Geographic Distribution of Animals." Wiley, New York.

Darwin, C. (1859). "On the Origin of the Species by Means of Natural Selection, on the Preservation of Favored Races in the Struggle for Life." Murray, London.

Day, J. H., Field, J. G., and Montgomery, M. P. (1971). The use of numerical methods to determine the distribution of the benthic fauna across the continental shelf off North Carolina. *J. Anim. Ecol.* **40,** 93–123.

Delany, M. J., and Healy, M. J. R. (1964). Variation in the long-tailed fieldmouse [*Apodemus sylvaticus* (L)] in northwest Scotland. II. Simultaneous examination of all characters. *Proc. Roy. Soc., Ser. B* **161,** 200–207.

Dickman, M. (1968). Some indices of diversity. *Ecology* **49,** 1191–1193.

Ducker, S. C., Williams, W. T., and Lance, G. N. (1965). Numerical classification of the Pacific forms of *Chlorodesmis* (Chlorophyta). *Aust. J. Bot.* **13,** 489–499.

Dunbar, C. O. (1950). The species concept: Further discussion. *Evolution* **4,** 175–176.

Ebeling, A. W., Ibara, R. M., Lavenberg, R. J., and Rohlf, F. J. (1970). Ecological groups of deep-sea animals off Southern California. *Bull. Los Angeles County Mus. Natur. Hist. Sci.* **6,** 1–43.

Edden, A. C. (1971). A measure of species diversity related to the lognormal distribution of individuals among species. *J. Exp. Mar. Biol. Ecol.* **6,** 199–209.

Edye, L. A., Williams, W. T., and Pritchard, A. J. (1970). A numerical analysis of variation patterns in Australian introductions of *Glycine wightii* (*G. javanica*). *Aust. J. Agr. Res.* **71,** 57–69.

Ehrlich, P. R. (1961). Systematics in 1970: Some unpopular predictions. *Syst. Zool.* **10,** 167–176.

Ekman, S. (1953). "Zoogeography of the Sea." Sidgwick & Jackson, London.

El-Gazzar, A., and Watson, L. (1970). A taxonomic study of Labiatae and related genera. *New Phytol.* **69,** 451–486.

El-Gazzar, A., Watson, L., Williams, W. T., and Lance, G. N. (1968). The taxonomy of *Salvia*: A test of two radically different numerical methods. *J. Linn. Soc. London, Bot.* **60,** 237–250.

Elton, C. (1947). "Animal Ecology." Sidgwick & Jackson, London.

Emerson, A. (1939). Social co-ordination and the super-organism. *Proc. Conf. Amer. Midl. Natur.* **21,** 182–209.

Estabrook, G. F. (1967). An information theory model for character analysis. *Taxon* **16,** 86–97.

Fager, E. W. (1957). Determination and analysis of recurrent groups. *Ecology* **38,** 586–593.

Fager, E. W. (1963). Communities of organisms. *In* "The Sea. Ideas and Observations on Progress in the Study of the Seas" (M. N. Hill, ed.), pp. 415–433. Wiley (Interscience), New York.

Fager, E. W. (1968). The community of invertebrates in decaying oak wood. *J. Anim. Ecol.* **37,** 121–142.

Fager, E. W., and Longhurst, A. R. (1968). Recurrent group analysis of species assemblages of demersal fish in the Gulf of Guinea. *J. Fish. Res. Bd. Can.* **25,** 1405–1421.

Fager, E. W., and McGowan, J. (1963). Zooplankton species groups in the North Pacific. *Science* **140,** 453–460.

Fein, L. (1961). The computer-related sciences (synnoetics) at a university in the year 1975. *Amer. Sci.* **49,** 149–169.

Field, J. G. (1969). The use of the information statistic in the numerical classification of heterogeneous systems. *J. Ecol.* **57,** 565–569.

Field, J. G. (1970). The use of numerical methods to determine benthic distribution patterns from dredgings in False Bay. *Trans. Roy. Soc. S. Afr.* **39,** 183–200.

Field, J. G. (1971). A numerical analysis of changes in the soft-bottom fauna along a transect across False Bay, South Africa. *J. Exp. Mar. Biol. Ecol.* **7,** 215–253.

Field, J. G., and Macfarlane, G. (1968). Numerical methods in marine ecology. I. A quantitative "similarity" analysis of rocky shore samples in False Bay, South Africa. *Zool. Afr.* **3,** 119–137.

Field, J. G., and Robb, R. T. (1970). Numerical methods in marine ecology. II. Gradient analysis of rocky shore samples from False Bay. *Zool. Afr.* **5,** 191–210.

Fischer, A. G. (1960). Latitudinal variations in organic diversity. *Evolution* **14,** 64–81.

Fisher, L., and van Ness, J. W. (1971). Admissable clustering procedures. *Biometrika* **58,** 91–104.

Fisher, R. A., Corbet, A. S., and Williams, C. B. (1943). The relation between the number of individuals and the number of species in a random sample of an animal population. *J. Anim. Ecol.* **12,** 42–58.

Forbes, S. A. (1887). The lake as a microcosm. *Bull. Sci. A. Peoria*; reprinted in *Ill., State Natur. Hist. Surv., Bull.* **5,** 537–550 (1925).

Forbes, S. A. (1907). An ornithological cross-section of Illinois in autumn. *Ill., State Lab. Natur. Hist., Bull.* **7,** 305–335.

Fraser, J. H. (1952). The Chaetognatha and other zooplankton of the Scottish area and their value as biological indicators of hydrographical conditions. *Mar. Res. Scot.* **2,** 1–52.

Fraser, J. H. (1955). The plankton of the waters approaching the British Isles in 1953. *Mar. Res. Scot.* **1,** 1–12.

Gage, J. (1972). A preliminary survey of the benthic macrofauna and sediments in Lochs Etive and Creran, sea-lochs along the west coast of Scotland. *J. Mar. Biol. Ass. U.K.* **52,** 237–276.

George, T. N. (1956). Conclusion. Biospecies, chromospecies, and morphospecies. *In* "The Species Concept in Paleontology" (P. C. Sylvester-Bradley, ed.), pp. 123–137. Systematics Association, London.

Gleason, H. A. (1922). On the relation between species and area. *Ecology* **3,** 158–162.

Good, I. J. (1965). Categorization of classification. *In* "Mathematics and Computer Science in Medicine and Biology," pp. 115–128. HMS Stationery Office, London.

Goodall, D. W. (1953). Objective methods for the classification of vegetation. I. The use of positive interspecific correlation. *Aust. J. Bot.* **1,** 39–63.

Goodall, D. W. (1954). Objective methods in the classification of vegetation. III. An essay in the use of factor analysis. *Aust. J. Bot.* **2,** 303–324.

Goodall, D. W. (1966a). A new similarity index based on probability. *Biometrics* **22,** 882–907.

Goodall, D. W. (1966b). Numerical taxonomy of bacteria—some published data re-examined. *J. Gen. Microbiol.* **42,** 25–37.

Goodman, L. A., and Kruskal, W. H. (1954). Measures of association for cross classifications. *J. Amer. Statist. Ass.* **49,** 732–764.

Goodman, L. A., and Kruskal, W. H. (1959). Measures of association for cross classification. *J. Amer. Statist. Ass.* **54,** 123–163.

Gould, S. W. (1958). Punched cards, binomial names and numbers. *Amer. J. Bot.* **45,** 331–339.

Gould, S. W. (1963). International plant index. Its methods, purposes and future possibilities. *Taxon* **12,** 177–182.

Goulden, C. E. (1966). The history of Laguna de Petenzil: The animal microfossils. *Mem. Conn. Acad. Arts & Sci.* **17,** 84–120.

Gower, J. C. (1966). Some distance properties of latent root and vector methods used in multivariate analysis. *Biometrika* **53,** 325–338.

Gower, J. C. (1967). Multivariate analysis and multidimensional geometry. *Statistician* **17,** 13–28.

Gower, J. C. (1969). A survey of numerical methods useful in taxonomy. *Acaralogia* **11,** 357–375.

Gower, J. C., and Ross, G. J. S. (1969). Minimum spanning trees and single linkage cluster analysis. *Appl. Statist.* **18,** 54–64.

Greene, E. L. (1909). Landmarks of botanical history. *Smithson. Misc. Collect.* **54,** 1–329.

Greig-Smith, P. (1964). "Quantitative Plant Ecology," 2nd ed. Butterworth, London.

Greig-Smith, P. (1971). Analysis of vegetation data: The user viewpoint. *In* "Statistical Ecology" (G. P. Patil, E. C. Pielou, and W. E. Waters, eds.), Vol. 3, pp. 149–166. Penn. State Univ. Press, University Park, Pennsylvania.

Greig-Smith, P., Austin, M. P., and Whitmore, T. C. (1967). The application of quantitative methods to vegetation survey. I. Association-analysis and principle component ordination of rain forest. *J. Ecol.* **55,** 483–503.

Griffin, D. J. G., and Yaldwyn, J. C. (1968). The constitution, distribution and relationships of the Australian decapod Crustacea. *Proc. Linn. Soc. N.S.W.* **93,** 165–183.

Grisebach, A. H. R. (1838). Ueber den Einfluss des Klimas auf die Begrenzung der natürlichen Floren. *Linnaea* **12,** 159–200.

Haeckel, E. (1866). "Generelle Morphologie der Organisimen.," Vol. II. Georg Reimen, Berlin.

Hagmeier, E. M. (1966). Numerical analysis of the distributional patterns of North American mammals. 2. Re-evaluation of the provinces. *Syst. Zool.* **15,** 279–299.

Hagmeier, E. M., and Stults, C. D. (1964). A numerical analysis of the distributional patterns of north American mammals. *Syst. Zool.* **13,** 125–155.

Hailstone, T. H. (1972). Ecological studies on the sub-tidal benthic macrofauna at the mouth of the Brisbane River. Ph.D. Thesis, University of Queensland, Brisbane, Australia.

Haldane, J. B. S. (1956). Can a species concept by justified? *In* "The Species Concept in Paleontology" (P. C. Sylvester-Bradley, ed.), pp. 95–96. Systematics Association, London.

Hamann, U. (1961). Merkmalsbestand und Verwandtschaftsbeziehungen den Farinosae. Ein Beitrag zum system der Monokotyledonen. *Willdenowia* **2,** 639–768.

Harman, H. H. (1968). "Modern Factor Analysis." Univ. of Chicago Press, Chicago, Illinois.

Hawkesworth, F. G., Estabrook, G. F., and Rogers, D. J. (1968). Application of an information theory model for character analysis in the genus *Arceuthobium* (Viscaceae). *Taxon* **17,** 605–608.

Heincke, F. (1898). Naturgeschichte der Herings. I. Die Lokalformen und die Wanderungen des Herings in den europaischen Meeren. *Abh. Deut. Seefischerei Ver.* **2,** 1–223.

Hemmings, S. K., and Rostron, J. (1972). A multivariate analysis on the Scottish Middle Old Red Sandstone antiarch fish genus *Pterichthyodes* Bleeker. *Biol. J. Linn. Soc.* **4,** 15–28.

Hendrickson, J. A., Jr., and Ehrlich, P. R. (1971). An expanded concept of "species diversity." *Notulae Natur. Acad. Natur. Sci. Philadelphia* **439,** 1–6.

Hensen, V. (1887). Ueber die Bestimmung des Planktons oder des in Meere treibendon Materials an Pflanzen und Thieren. *Ber. Komm. Wiss. Unters. Deut. Meere, Kiel* **5,** 1–107.

Hesse, R. (1924). "Tiergeographie aud Okologischer Grundlage." Fischer, Jena.

Hesse, R., Allee, W. C., and Schmidt, K. P. (1937). "Ecological Animal Geography." Wiley, New York (fourth reprinting, 1949).

Hessler, R. R., and Sanders, H. L. (1967). Faunal diversity in the deep sea. *Deep Sea Res.* **14,** 65–78.

Holloway, J. D., and Jardine, N. (1968). Two approaches to zoogeography: A study based on the distributions of butterflies, birds and bats in the Indo-Australian area. *Proc. Linn. Soc. London* **179,** 153–188.

Holme, N. A. (1949). A new bottom-sampler. *J. Mar. Biol. Ass. U.K.* **28,** 323–332.

Holme, N. A. (1951). Sampling the sea-bed. *Discovery* **12,** 59–63.

Holme, N. A. (1953). The biomass of the bottom fauna in the English Channel off Plymouth. *J. Mar. Biol. Ass. U.K.* **32,** 1–49.

Holme, N. A. (1955). An improved "vacuum" grab for sampling the sea-floor. *J. Mar. Biol. Ass. U.K.* **38,** 545–551.

Holme, N. A. (1964). Methods of sampling the benthos. *Advan. Mar. Biol.* **2,** 171–260.

Hope-Simpson, J. F. (1940). On the errors in the ordinary use of subjective frequency estimates in grassland. *J. Ecol.* **28,** 193–209.

Hughes, R. N., and Thomas, M. L. H. (1971a). The classification and ordination of shallow-water benthic samples from Prince Edward Island, Canada. *J. Exp. Mar. Biol. Ecol.* **7,** 1–39.

Hughes, R. N., and Thomas, M. L. H. (1971b). Classification and ordination of benthic samples from Bedeque Bay, an estuary in Prince Edward Island, Canada. *Mar. Biol.* **10,** 227–235.

Huheey, J. E. (1966). A mathematical methods of analysing biogeographical data. I. Herpertofauna of Illinois. *Amer. Midl. Natur.* **73,** 490–500.

Hurlbert, S. H. (1971). The non-concept of species diversity: A critique and alternative parameters. *Ecology* **52,** 557–586.

Hutchinson, G. E. (1957). Concluding remarks. *Cold Spring Harbor Symp. Quant. Biol.* **22,** 415–427.

Hutchinson, G. E. (1959). Homage to Santa Rosalia, or, Why are there so many kinds of animals? *Amer. Natur.* **93,** 145–159.

Huxley, J. S. (Ed.). (1940, reprint 1949). "The New Systematics." Oxford Univ. Press, London and New York.

Huxley, J. S. (1942). "Evolution: The Modern Synthesis." Allen & Unwin, London.

Ivimey-Cook, R. B., Proctor, M. C. F., and Wigston, D. L. (1969). On the problem of the 'R/Q' terminology in multivariate analyses of biological data. *J. Ecol.* **57,** 673–675.

Jaccard, P. (1908). Nouvelles recherches sur la distribution florale. *Bull. Soc. Vaud. Sci. Natur.* **44,** 223–270.

Jahn, T. L. (1961). Man versus machine: A future problem in protozoan taxonomy. *Syst. Zool.* **10,** 179–192.

Jamieson, B. G. M. (1968). A taxonomic investigation of the Alluroididae (Oligochaeta). *J. Zool.* **155,** 55–86.

Jancey, R. C. (1966a). Multidimensional group analysis. *Aust. J. Bot.* **14,** 127–130.

Jancey, R. C. (1966b). The application of numerical methods of data analysis to the genus *Phyllota* Benth. in New South Wales. *Aust. J. Bot.* **14,** 131–149.

Jardine, N., and Sibson, R. (1968). The construction of hierarchic and non-hierarchic classifications. *Comput. J.* **11,** 177–184.

Jardine N., and Sibson, R. (1971a). Choice of methods for automatic classification. *Comput. J.* **14,** 404–406.

Jardine, N., and Sibson, R. (1971b). "Mathematical Taxonomy." Wiley, New York.

Jones, G. F. (1969). The benthic macrofauna of the mainland shelf of southern California. *Allan Hancock Monogr. Mar. Biol.* **4,** 1–219.

Kelley, J. L. (1955). "General Topology." Van Nostrand, New York.

Kendall, M. G. (1966). Discrimination and classification. *In* "Multivariate Analysis" (P. R. Krishnaiah, ed.), pp. 165–185. Academic Press, New York.

Kendall, M. G., and Stuart, A. (1966). "The Advanced Theory of Statistics," 3 vols. Griffin, London.

Kendrick, W. B. (1965). Complexity and dependence in computer taxonomy. *Taxon* **14,** 141–154.

Kikkawa, J. (1968). Ecological association of bird species and habitats in eastern Australia; similarity analysis. *J. Anim. Ecol.* **37,** 143–165.

Kikkawa, J., and Pearse, K. (1969). Geographical distribution of land birds in Australia—a numerical analysis. *Aust. J. Zool.* **17,** 821–840.

Klauber, L. M. (1940). Two new subspecies of *Phyllorhynchus*, the leaf-nosed snake, with notes on the genus. *Trans. San Diego Soc. Natur. Hist.* **9,** 195–214.

Klopfer, P. H., and MacArthur, R. H. (1961). On the causes of tropical species diversity: Niche overlap. *Amer. Natur.* **95,** 223–226.

Knox, G. A. (1963). The biogeography and intertidal ecology of the Australasian coasts. *Oceanogr. Mar. Biol. Annu. Rev.* **1,** 341–404.

Kückler, A. W. (1967). "Vegetation Mapping." Ronald Press, New York.

Kulczynski, S. (1927). Die Pflanzenassociationen der Pienenen. *Bull. Int. Acad. Pol. Sci. Lett., Cl. Sci. Math. Natur., Ser. B* Suppl. **2,** 57–203.

Lam, H. J. (1936). Phylogenetic symbols, past and present. (Being an apology for genealogical trees.) *Acta Biotheor.* **2,** 153–193.

de Lamarck, J. B. M. (1809). "Philosophie zoologique." Dentu, Paris.

Lambert, J. M., and Dale, M. B. (1964). The use of statistics in phytosociology. *Advan. Ecol. Res.* **2,** 59–99.

Lance, G. N., and Williams, W. T. (1966a). Computer programs for hierarchical polythetic classification ("similarity analyses"). *Comput. J.* **9,** 60–64.

Lance, G. N., and Williams, W. T. (1966b). A generalised sorting strategy for computer classifications. *Nature* (*London*) **212,** 218.

Lance, G. N., and Williams, W. T. (1967a). A general theory of classificatory sorting strategies. I. Hierarchical systems. *Comput. J.* **9,** 373–380.

Lance, G. N., and Williams, W. T. (1967b). Mixed-data classificatory programs. I. Agglomerative systems. *Aust. Comput. J.* **1,** 15–20.

Lance, G. N., and Williams, W. T. (1968a). Mixed-data classificatory programs. II. Divisive systems. *Aust. Comput. J.* **1,** 82–85.

Lance, G. N., and Williams, W. T. (1968b). Note on a new information-statistic classificatory program. *Comput. J.* **11,** 195.

Lance, G. N., and Williams, W. T. (1971). A note on a new divisive classificatory program for mixed data. *Comput. J.* **14,** 154–155.

Lankester, E. R. (1889). Zoology. *In* "Encyclopedia Britannica," 9th ed., Vol. 24, pp. 799–820.

Lavarack, P. S. (1972). The taxonomic affinities of the Neottioideae." Ph.D. Thesis, University of Queensland, Brisbane, Australia.

Lie, U., and Kelley, J. C. (1970). Benthic infauna communities off the coast of Washington and in Puget Sound: Identification and distribution of the communities. *J. Fish. Res. Bd. Can.* **27,** 621–651.

Linnaeus, C. (1737). "Critica botanica." Lugduni, Batavorum.

Lloyd, M. (1967). Mean crowding. *J. Anim. Ecol.* **36,** 1–30.

Lloyd, M., and Ghelardi, R. J. (1964). A table for calculating the "equitability" component of species diversity. *J. Anim. Ecol.* **33,** 217–225.

Lloyd, M., Zar, J. H., and Karr, J. R. (1968). On the calculation of information-theoretical measures of diversity. *Amer. Midl. Natur.* **79,** 257–272.

Longhurst, A. R. (1969). Species assemblages in tropical demersal fisheries. *Proc. Symp. Oceanogr. Fish. Resour. Trop. Atl. UNESCO* pp. 147–168.

Loya, Y. (1972). Community structure and species diversity of hermatypic corals at Eilat, Red Sea. *Mar. Biol.* **13,** 100–123.

MacArthur, R. H. (1955). Fluctuations of animal populations, and a measure of community stability. *Ecology* **36,** 533–536.

MacArthur, R. H. (1957). On the relative abundance of bird species. *Proc. Nat. Acad. Sci. U.S.* **43,** 293–295.

MacArthur, R. H. (1960). On the relative abundance of species. *Amer. Natur.* **94,** 25–36.

MacArthur, R. H. (1965). Patterns of species diversity. *Biol. Rev., Cambridge Phil. Soc.* **40,** 510–533.

MacArthur, R. H. (1969). Patterns of communities in the tropics. *Biol. J. Linn. Soc.* **1,** 19–30.

MacArthur, R. H., and MacArthur, J. W. (1961). On bird species diversity. *Ecology* **42,** 594–598.

MacArthur, R. H., and Wilson, E. O. (1967). "The Theory of Island Biogeography." Univ. of Princeton Press, Princeton, New Jersey.

Macfadyen, M. A. (1963). "Animal Ecology; Aims and Methods." Pitman, London.

McConnaughey, B. H. (1964). The determination and analysis of plankton communities. *Mar. Res. Indonesia Spec.* pp. 1–40.

McIntosh, R. P. (1967a). An index of diversity and the relation of certain concepts to diversity. *Ecology* **48,** 392–404.

McIntosh, R. P. (1967b). The continuum concept of vegetation. *Bot. Rev.* **33,** 130–187.

Macnaughton-Smith, P. (1965). "Some Statistical and other Numerical Techniques for Classifying Individuals. Home Office. Studies in the Causes of Delinquency and the Treatment of Offenders," No. 6. HM Stationery Office, London.

Macnaughton-Smith, P., Williams, W. T., Dale, M. B., and Mockett, L. G. (1964). Dissimilarity analysis: A new technique of hierarchical subdivision. *Nature (London)* **202,** 1034–1035.

McNeill, J. (1972). The hierarchical ordering of characters as a solution to the character problem in numerical taxonomy. *Taxon* **21,** 71–82.

Malthus, T. R. (1798). "An Essay on the Principle of Population." Johnston, London (reprinted: Everyman's Library, 1914).

Mankowski, W. (1962). Biological macroplankton indicators of the inflow of salt water from the North Sea into the Baltic Sea. *Przegl. Zool.* **6,** 38–42.

Margalef, R. (1951). Diversidad de especies en les communidades naturales. *Publ. Inst. Biol. Apl., Barcelona* **6,** 59–72.

Margalef, R. (1957). La teoria de la informacion en ecologia. *Mem. Real. Acad. Cienc. Artes Barcelona* [3] **32,** 373–449, translation: Information theory in ecology. *Gen. Syst.* **3,** 36–71 (1958).

Margalef, R. (1968). "Perspectives in Ecological Theory." Univ. of Chicago Press, Chicago, Illinois.

Martin, W. E., Duke, J. A., Bloom, S. G., and McGinnis, J. T. (1970). Possible effects of a sea-level canal on the marine ecology of the American Isthmian region. *Battelle Mem. Inst., Ohio* pp. 1–220.

Mayr, E. (1949). The species concept: Semantics versus semantics. *Evolution* **3,** 371–374.

Mayr, E., Linsley, E. G., and Usinger, R. L. (1953). "Methods and Principles of Systematic Zoology." McGraw-Hill, New York.

Michener, C. D., and Sokal, R. R. (1957). A quantitative approach to a problem in classification. *Evolution* **11,** 130–162.

Mills, E. L. (1969). The community concept in marine zoology, with comments on continua and instability in some marine communities: A review. *J. Fish. Res. Bd. Can.* **26,** 1415–1428.

Milne, M. J., and Milne, L. J. (1939). Evolutionary trends in caddis worm case construction. *Ann. Entomol. Soc. Amer.* **32,** 533–542.

Möbius, K. (1877). "Die Auster und die Austerwirtschaft." Parey, Berlin.

Monk, C. D. (1967). Tree species diversity in the eastern deciduous forest with particular reference to North Central Florida. *Amer. Natur.* **100,** 65–75.

Morrison, D. F. (1967). "Multivariate Statistical Methods." McGraw–Hill, New York.

Motyka, J., Dobrzanski, B., and Zawazki, S. (1950). Wstspne badania nad lagami poludnio-wowschodniez Lubelszczyzny. (Preliminary studies on meadows in the southeast of the province Lublin.) *Univ. Mariae Curie-Sklodowska Ann., Sect. E* **5,** 367–447.

Needham, R. M. (1962). A method for using computers in information classification. *Proc. Int. Fed. Information Processing Soc. Congr.* **62,** 284–287.

Needham, R. M., and Jones, K. S. (1964). Keywords and clumps. *J. Doc.* **20,** 5–15.

Nordenskiöld, E. (1928). "The History of Biology: A Survey." Knopf, New York.

Noy-Meir, I. (1970). Component analysis of semi-arid vegetation in southeastern Australia. Ph.D. Thesis, Australian National University, Canberra, Australia.

Noy-Meir, I. (1971). Multivariate analysis of the semi-arid vegetation in south-eastern Australia. Nodal ordination by component analysis. *Quantifying Ecol., Proc. Ecol. Soc. Aust.* **6,** 159–193.

Noy-Meir, I. (1973). Data transformations in ecological ordination. I. Some advantages of non-centering. *J. Ecol.* **61,** 329–341.

Noy-Meir, I., and Austin, M. P. (1970). Principal component ordination and simulated vegetation data. *Ecology* **51,** 551–552.

Ochiai, A. (1957). Zoogeographic studies on the soleoid fishes found in Japan and its neighboring regions. (In Japanese, English summary.) *Bull. Jap. Soc. Sci. Fish.* **22,** 526–530.

Odum, E. P. (1971). "Fundamentals of Ecology." Saunders, Philadelphia, Pennsylvania.

Odum, H. T. (1967). Biological circuits and the marine systems of Texas. *In* "Pollution and Marine Biology" (T. A. Olson and F. J. Burgess, eds.), pp. 99–157. Wiley (Interscience), New York.

Orloci, L. (1966). Geometric models in ecology. I. The theory and application of some ordination methods. *J. Ecol.* **54,** 193–215.

Orloci, L. (1967a). Data centering: A review and evaluation with reference to component analysis. *Syst. Zool.* **16,** 208–212.

Orloci, L. (1967b). An agglomerative method for classification of plant communities. *J. Ecol.* **55,** 193–206.

Orloci, L. (1969). Information analysis of structure in biological collections. *Nature (London)* **223,** 483–484.

Patrick, R. (1949). A proposed biological measure of stream conditions. *Proc. Acad. Natur. Sci. Philadelphia* **101,** 277–341.

Patrick, R., and Strawbridge, D. (1963). Variation in the structure of natural diatom communities. *Amer. Natur.* **98,** 51–57.

Patrick, R., Hohn, M. H., and Wallace, J. H. (1954). A new method for determining the pattern of the diatom flora. *Notulae Natur. Acad. Natur. Sci., Philadelphia* **254,** 1–12.

Patten, B. C. (1962). Species diversity in net phytoplankton of Raritan Bay. *J. Mar. Res.* **20,** 57–75.

Pearson, T. H. (1970). The benthic ecology of Loch Linnhe and Loch Eil, a sea-loch system on the west coast of Scotland. I. The physical environment and the distribution of the macro-benthic fauna. *J. Exp. Mar. Biol. Ecol.* **5,** 1–34.

Pearson, T. H. (1971). The benthic ecology of Loch Linnhe and Loch Eil, a sea-loch system on the west coast of Scotland. III. The effect on the benthic fauna of the introduction of pulp mill effluent. *J. Exp. Mar. Biol. Ecol.* **6,** 211–233.

Pérès, J. M., and Picard, J. (1958). "Manuel de bionomie benthique de la mer Méditerranée." Louis-Jean, Gap, France.

Peters, J. A. (1971). A new approach in the analysis of biogeographic data. *Smithson. Contrib. Zool.* **107,** 1–28.

Petersen, C. G. J. (1914). Valuation of the sea. II. The animal communities of the sea bottom and their importance for marine zoogeography. *Rep. Dan. Biol. Sta.* **21,** 1–44. (Appendix).

Pianka, E. R. (1966). Latitudinal gradients in species diversity: A review of concepts. *Amer. Natur.* **100,** 33–46.

Pielou, E. C. (1966). Species-diversity and pattern-diversity in the study of ecological succession. *J. Theor. Biol.* **10,** 370–383.

Pielou, E. C. (1967). The use of information theory in the study of the diversity of biological populations. *Proc. Berkeley Symp. Math. Statist. Probab., 5th,* Vol. 4, pp. 163–177.

Pielou, E. C. (1969). "An Introduction to Mathematical Ecology." Wiley (Interscience), New York.

Pielou, E. C. (1972). Niche width and niche overlap: A method for measuring them. *Ecology* **53,** 687–692.

Poore, M. E. D. (1955a). The use of phytosociological methods in ecological investigations. I. The Braun-Blanquet system. *J. Ecol.* **43,** 226–244.

Poore, M. E. D. (1955b). The use of phytosociological methods in ecological investigations. II. Practical issues involved in an attempt to apply the Braun-Blanquet system. *J. Ecol.* **43,** 245–269.

Poore, M. E. D. (1955c). The use of phytosociological methods in ecological investigations. III. Practical applications. *J. Ecol.* **43,** 606–651.

Popham, J. D., and Ellis, D. V. (1971). A comparison of traditional, cluster and Zürich-Montpellier analyses of infaunal pelecypod associations from two adjacent sediment beds. *Mar. Biol.* **8,** 260–266.

Preston, F. W. (1948). The commonness, and rarity of species. *Ecology* **29,** 254–283.

Preston, F. W. (1962). The canonical distribution of commonness and rarity. *Ecology* **43,** 185–216 and 410–432.

Pritchard, D. W. (1967). What is an estuary: Physical viewpoint. *In* "Estuaries," Publ. No. 83, pp. 3–5. Amer. Ass. Advan. Sci., Washington, D.C.

Pritchard, N. M., and Anderson, A. J. B. (1971). Observations on the use of cluster analysis in botany with an ecological example. *J. Ecol.* **59,** 727–747.

Pulliam, H. R., Odum, E. P., and Barrett, G. W. (1968). Equitability and resource limitation. *Ecology* **49,** 772–774.

Ramaley, F. (1940). The growth of a science. *Univ. Colo. Stud., Gen. Ser.* **26,** 3–14.

Rao, C. R. (1952). "Advanced Statistical Methods in Biometric Research." Wiley, New York.

Raphael, Y. I., and Stephenson, W. (1972). "The Macrobenthos of Bramble Bay, Moreton Bay, Southern Queensland," Cyclostyled report. Queensland Dept. Coordinator General and Commonwealth Dept. Works and Housing, Melbourne, Australia.

Raunkiaer, C. (1934). "The Life Forms of Plants and Statistical Plant Geography, Being the Collected Papers of C. Raunkier" (A. G. Tansley, ed.), pp. 1–632. Oxford Univ. Press (Clarenden), London and New York.

Resvoy, P. D. (1924). Zur definition des Biocönose—Begriffs. *Russ. Gidrobiol. Zh.* **3,** 204–209.

Rhodes, A. M., Campbell, C., Malo, S. E., and Carmer, S. G. (1970). A numerical taxonomic study of mango, *Mangifera indica*. *J. Amer. Soc. Hort. Sci.* **95,** 252–256.

Rhodes, A. M., Malo, S. E., Campbell, C. W., and Carmer, S. G. (1971). A numerical taxonomic study of the avocado (*Persea americana* Mill.). *J. Amer. Soc. Hort. Sci.* **96,** 391–395.

Rhodes, F. H. T. (1956). The time factor in taxonomy. *In* "The Species Concept in Paleontology" (P. C. Sylvester-Bradley, ed.), pp. 33–52. Systematics Association, London.

Rogers, D. J., and Fleming, H. (1964). A computer program for classifying plants. II. *Bioscience* **14,** 15.

Rogers, D. J., and Tanimoto, T. T. (1960). A computer program for classifying plants. *Science* **132,** 1115–1118.

Rostron, J. (1972). A multivariate statistical study of skull measurements of five taxa of gazelles. *Biol. J. Linn. Soc.* **4,** 1–14.

Russell, F. S. (1935). On the value of certain plankton animals as indicators of water movements in the English Channel and North Sea. *J. Mar. Biol. Ass. U.K.* **20,** 309–332.

Russell, F. S. (1939). Hydrographical and biological conditions in the North Sea as indicated by plankton organisms. *J. Cons., Cons. Perm. Int. Explor. Mer.* **14,** 171–192.

Russell, P. F., and Rao, T. R. (1940). On habitat and association of species of anopheline larvae in south-eastern Madras. *J. Malaria Inst. India* **3,** 153–178.

Sager, P. E., and Hasler, A. D. (1969). Species diversity in lacustrine phytoplankton. I. The components of the index of diversity from Shannon's formula. *Amer. Natur.* **103,** 51–59.

San, M. M. (1971). The genus *Lomandra* in eastern Australia, a biological and numerical taxonomic study. Ph.D. thesis, University of Queensland, Brisbane, Australia.

Sanders, H. L. (1968). Marine benthic diversity: A comparative study. *Amer. Natur.* **102,** 243–282.

Sands, W. A. (1972). The soldierless termites of Africa (Isoptera: Termitidae). *Bull. Brit. Mus. Natur. Hist.* **18,** (Ent.) Suppl., 3–243.

Savory, T. (1970). "Animal Taxonomy." Heinemann, London.

Seal, H. L. (1964). "Multivariate Statistical Analysis for Biologists." Metheun, London.

Shannon, C. E. (1948). A mathematical theory of communication. *Bull. Syst. Tech. J.* **27,** 379–423 and 623–656.

Shannon, C. E., and Weaver, W. (1963). "The Mathematical Theory of Communication." Univ. of Illinois Press, Urbana.

Sheals, J. G. (1965). The application of computer techniques to acarine taxonomy: A preliminary examination with species of *Hypoaspis-Androlaelaps* complex (Acarina). *Proc. Linn. Soc. London* **176,** 11–21.

Sheard, K. (1965). Species groups in the zooplankton of the eastern Australian slope waters, 1938–1941. *Aust. J. Mar. Freshwater Res.* **16,** 219–254.

Shelford, V. E. (1907). Preliminary notes on the distribution of Tiger beetles (*Cicindela*) and its relation to plant succession. *Biol. Bull.* **21,** 9–34.

Sibson, R. (1971). Some observations of a paper by Lance and Williams. *Comput. J.* **14,** 156–157.

Siegel, S. (1956). "Nonparametric Statistics for the Behavioural Sciences." McGraw-Hill, New York.

Simon, F. H. (1971). Prediction methods in criminology. "Home Office Research Studies," Vol. 7. HM Stationery Office, London.

Simpson, E. H. (1949). Measurement of diversity. *Nature* (*London*) **163,** 688.

Simpson, G. G. (1960). Notes on the measurement of faunal resemblance. *Amer. J. Sci., Bradley Vol.* **258-A,** 300–311.

Simpson, G. G. (1961). "Principles of Animal Taxonomy." Columbia Univ. Press, New York.

Sims, R. W. (1966). The classification of the megascolecoid earthworms: An investigation of oligochaete systematics by computer techniques. *Proc. Linn. Soc. London* **177,** 125–141.

Smith, A. D. (1944). A study of the reliability of range vegetation estimates. *Ecology* **25,** 441–448.

Sneath, P. H. A., and Sokal, R. R. (1973). "Numerical Taxonomy. The Principles and Practice of Numerical Classification." Freeman, San Francisco, California.

Sokal, R. R., and Michener, C. D. (1958). A statistical method for evaluating systematic relationships. *Univ. Kans. Sci. Bull.* **38,** 1409–1438.

Sokal, R. R., and Rohlf, F. J. (1962). The comparison of dendrograms by objective methods. *Taxon* **11,** 33–40.

Sokal, R. R., and Sneath, P. H. A. (1963). "Principles of Numerical Taxonomy." Freeman, San Francisco, California.

Sørensen, T. (1948). A method of establishing groups of equal amplitude in plant sociology based on similarity of species content and its application to analyses of the vegetation on Danish commons. *Biol. Skr.* **5,** 1–34.

Southward, A. J. (1962). The distribution of some plankton animals in the English Channel and approaches. II. *J. Mar. Biol. Ass. U.K.* **42,** 275–375.

Southward, A. J. (1967). Recent changes in abundance of intertidal barnacles in South-West England. *J. Mar. Biol. Ass. U.K.* **47,** 81–95.

Spurway, H. (1955). The subhuman capacities for species recognition and their correlation with reproductive isolation. *Proc. Int. Ornithol. Congr., 11th, 1954* (quoted from Haldane, 1956).

Stearn, W. T. (1959). The background of Linnaeus' contributions to the nomenclature and methods of systematic biology. *Syst. Zool.* **8,** 4–22.

Stenzel, H. B. (1949). Successional speciation in paleontology: The case of the oysters. *Evolution* **3,** 34–50.

Stephenson, W. (1972). An annotated check list and key to the Indo-West-Pacific swimming crabs (Crustacea: Decapoda: Portunidae). *Bull. Roy. Soc. N.Z.* **10,** 1–62.

Stephenson, W. (1973a). The use of computers in classifying marine bottom communities. "Oceanography South Pacific 1972" (compiled by R. Fraser). pp. 463–473 N.Z. Nat. Comm. for UNESCO, Wellington.

Stephenson, W. (1973b). The validity of the community concept in marine biology. *Proc. Roy. Soc. Queensl.* **84,** 73–86.

Stephenson, W., and Williams, W. T. (1971). A study of the benthos of soft bottoms, Sek Harbour, New Guinea, using numerical analysis. *Aust. J. Mar. Freshwater Res.* **22,** 11–34.

Stephenson, W., Williams, W. T., and Lance, G. N. (1968). Numerical approaches to the relationships of certain American swimming crabs (Crustacea: Portunidae). *Proc. U.S. Nat. Mus.* **124,** 1–25.

Stephenson, W., Williams, W. T., and Lance, G. N. (1970). The macrobenthos of Moreton Bay. *Ecol. Monogr.* **40,** 459–494.

Stephenson, W., Williams, W. T., and Cook, S. (1972). Computer analyses of Petersen's original data on bottom communities. *Ecol. Monogr.* **42,** 387–415.

Stephenson, W., Williams, W. T., and Cook, S. (1974). The macrobenthos of soft bottoms in southern Moreton Bay (south of Peel Island). *Mem. Queensl. Mus.* **17,** 73–124.

Strangeland, C. E. (1904). Pre-Malthusian doctrines of population. *Columbia Stud. Soc. Sci.: Stud. Hist. Econ. Soc. Law* **21,** 25.

Swan, J. M. A. (1970). An examination of some ordination problems by use of simulated vegetational data. *Ecology* **51,** 89–102.

Sylvester-Bradley, P. C. (1956). The new paleontology. *In* "The Species Concept in Paleontology" (P. C. Sylvester-Bradley, ed.), pp. 1–8. Systematics Association, London.

Taylor, L. R. (1961). Aggregation, variance and mean. *Nature* (*London*) **189,** 732–735.

Taylor, L. R. (1971). Aggregation as a species characteristic. *In* "Statistical Ecology" (G. P. Patel, E. C. Pielou, and W. E. Waters, eds.), Vol. I, pp. 356–377. Penn. State Univ. Press, University Park, Pennsylvania.

Tenore, K. R. (1972). Macrobenthos of the Pamlico River Estuary, North Carolina. *Ecol. Monogr.* **42,** 51–69.

Thomas, G. (1956). The species conflict-abstractions and their applicability. *In* "The Species Concept in Paleontology" (P. C. Sylvester-Bradley, ed.), pp. 17–31. Systematics Association, London.

Thorrington-Smith, M. (1971). West Indian Ocean phytoplankton: a numerical investigation of phyto-hydrographic regions and their characteristic phytoplankton associations. *Mar. Biol.* **9,** 115–137.

Thorson, G. (1952). Zur jetzigen Lage der marinen Bodentier-Ökologie. *Verh. Deut. Zool. Ges. Wilhelmshaven,* 1951, pp. 276–327.

't Mannetje, L. (1967a). A re-examination of the taxonomy of the genus *Rhizobium* and related genera using numerical analysis. *Antonie van Leeuwenhoek; J. Microbiol. Serol.* **33,** 477–491.

't Mannetje, L. (1967b). A comparison of eight numerical procedures applied to the classification of some African *Trifolium* taxa based on *Rhizobium* affinities. *Aust. J. Bot.* **15,** 521–528.

Tracey, J. G. (1968). Investigations of changes in pasture composition by some classificatory methods. *J. Appl. Ecol.* **5**, 639–648.

von Humboldt, A. (1805). "Essai sur la géographie des plantes; accompagné d'un tableau physique des régions équinociales" (by A. de Humboldt and A. Bonpland, edited by A. de Humboldt). Levrault, Schoell Co., Paris.

von Post, H. (1868). Försök till iakttagelser i djur-och växt-statistik. *Kgl. Vetenskarsakad. Foerh.* **24**, 35–79.

Wallace, C. C. (1972). An examination of the classification of some Australian megascolecoid earthworms (Annelida: Oligochaeta) by numerical methods. *Mem. Queensl. Mus.* **16**, 191–209.

Ward, J. H. (1963). Hierarchical grouping to optimize an objective function. *J. Amer. Statist. Ass.* **58**, 236–244.

Webb, L. J., Tracey, J. G., Williams, W. T., and Lance, G. N. (1967). Studies in the numerical analysis of complex rain-forest communities. I. A comparison of methods applicable to site/species data. *J. Ecol.* **55**, 171–191.

Webb, L. J., Tracey, J. G., Williams, W. T., and Lance, G. N. (1971). Prediction of agricultural potential from intact forest vegetation. *J. Appl. Ecol.* **8**, 99–121.

Webb, L. J., Tracey, J. G., Kikkawa, J., and Williams, W. T. (1973). Techniques for selecting and allocating land for nature conservation in Australia. *In* "Nature Conservation in the Pacific" (A. B. Costen, ed.), pp. 39–52. Aust. Nat. Univ. Press., Canberra, Australia.

Westoll, T. S. (1956). The nature of fossil species. *In* "The Species Concept in Paleontology" (P. C. Sylvester-Bradley, ed.), pp. 53–62. Systematics Association, London.

Whittaker, R. H. (1962). Classification of natural communities. *Bot. Rev.* **28**, 1–239.

Whittaker, R. H. (1965). Dominance and diversity in land plant communities. *Science* **147**, 250–260.

Whittaker, R. H. (1970). "Communities and Ecosystems." Macmillan, New York.

Whittaker, R. H. (1972). Evolution and measurement of species diversity. *Taxon* **21**, 213–251.

Whittaker, R. H., and Woodwell, G. M. (1969). Structure, production and diversity of the oak-pine forest at Brookhaven, New York. *J. Ecol.* **57**, 155–174.

Williams, W. T. (1969). The problem of attribute-weighting in numerical classification. *Taxon* **18**, 369–374.

Williams, W. T. (1971). Principles of clustering. *Annu. Rev. Ecol. Syst.* **2**, 303–326.

Williams, W. T. (1972). The problem of pattern. *Aust. Math. Teacher* **28**, 103–109.

Williams, W. T., and Clifford, H. T. (1971). On the comparison of two classifications of the same set of elements. *Taxon* **20**, 519–522.

Williams, W. T., and Dale, M. B. (1965). Fundamental problems in numerical taxonomy. *Advan. Bot. Res.* **2**, 35–68.

Williams, W. T., and Lambert, J. M. (1959). Multivariate methods in plant ecology. I. Association-analysis in plant communities. *J. Ecol.* **47**, 83–101.

Williams, W. T., and Lambert, J. M. (1960). Multivariate methods in plant ecology. II. The use of an electronic digital computer for association-analysis. *J. Ecol.* **48**, 689–710.

Williams, W. T., and Lambert, J. M. (1961). Multivariate methods in plant ecology. III. Inverse association-analysis. *J. Ecol.* **49**, 717–729.

Williams, W. T., and Lance, G. N. (1968). Choice of strategy in the analysis of complex data. *Statistician* **18**, 31–44.

Williams, W. T., and Stephenson, W. (1973). The analysis of three-dimensional data (sites × species × times) in marine ecology. *J. Exp. Mar. Biol. Ecol.* **11**, 207–227.

Williams, W. T., Lambert, J. M., and Lance, G. N. (1966). Multivariate methods in plant ecology. V. Similarity analyses and information-analysis. *J. Ecol.* **54**, 427–446.

Williams, W. T., Lance, G. N., Webb, L. J., Tracey, J. G., and Connell, J. H. (1969). Studies in the numerical analysis of complex rain-forest communities. IV. A method for the elucidation of small-scale forest pattern. *J. Ecol.* **57**, 635–654.

Williams, W. T., Clifford, H. T., and Lance, G. N. (1971a). Group-size dependence: A rationale for choice between numerical classifications. *Comput. J.* **14**, 157–162.

Williams, W. T., Lance, G. N., Dale, M. B., and Clifford, H. T. (1971b). Controversy concerning the criteria for taxonometric strategies. *Comput. J.* **14**, 162–165.

Williams, W. T., Lance, G. N., Webb, L. J., and Tracey, J. G. (1973). Studies in the numerical analysis of complex rain-forest communities. *J. Ecol.* **61**, 47–70.

Wirth, M., Estabrook, G. F., and Rogers, D. J. (1966). A graph theory model for systematic biology, with an example for the Oncidiinae (Orchidaceae). *Syst. Zool.* **15**, 59–69.

Wolff, T. (1970). The concept of the hadal or ultra-abyssal fauna. *Deep Sea Res.* **17**, 983–1003.

Woodwell, G. M. (1967). Radiation and the patterns of nature. *Science* **156**, 461–470.

Yarranton, G. A. (1967a). Principal components analysis of data from saxicolous bryophyte vegetation at Steps Bridge, Devon. I. A quantitative assessment of variation in the vegetation. *Can. J. Bot.* **45**, 93–115;

Yarranton, G. A. (1967b). Principal components analysis of data from saxicolous bryophyte vegetation at Steps Bridge, Devon. II. An experiment with heterogeneity. *Can. J. Bot.* **45**, 229–247.

Yarranton, G. A. (1967c). Principal components analysis of data from saxicolous bryophyte vegetation at Steps Bridge, Devon. III. Correlation of variation in the vegetation with environmental variables. *Can. J. Bot.* **45**, 249–258.

Yarranton, G. A., Beasleigh, W. J., Morrison, R. G., and Shafi, M. I. (1972). On the classification of phytosociological data into non-exclusive groups with a conjecture about determining the optimum number of groups in a classification. *Vegetatio* **24**, 1–12.

Young, D. J., and Watson, L. (1970). The classification of the dicotyledons: A study of the upper levels of the hierarchy. *Aust. J. Bot.* **18**, 387–433.

Young, D. K., and Rhodes, D. C. (1971). Animal-sediment relations in Cape Cod Bay, Massachusetts. I. A transect study. *Mar. Biol.* **11**, 242–254.

Subject Index

A

Abiotic, *see* Attribute
Abundance, 137, 147, 157, 164, *see also* Ubiquity
Acacia, 116, 178
Acorn barnacle, 42
Adansonian law, 14
Agglomerative, *see* Classification
Algae, 112, 183
Amoeba, 17
Apes, 64
Arbitrary, 5, 137, 147, 151
Arceuthobium, 206
Aristotle, 2, 4, 20, 21
Association, 16, 19, 21, 23, 50, 143, 149, 165
Association measures
 Chi-squared (χ^2), 57, 63, 77, 102, 103, 136, 155, 157
 coefficients, 50, 61–65
 correlation coefficient, 63, 64, 78, 87, 119, 142
  application to
   binary data, 62–63
   continuous data, 63–65
   meristic data, 63–65
 McConnaughy, 62
 mean square contingency, 63
 Pearson, 62
 Yule, 62
Attribute(s), 12, 34–35
 abiotic, 18, 19, 20, 34, 79, 90, 137, 143, 145, 155
 binary, 34, 36–41, 48, 52, 62, 72, 75, 105, 125, 127, 136, 146, 155, 157, 163, 195
 consequential, 36, 37 40
 continuous, 35, 39, 43, 46, 48, 63, 64, 75, 87, 105, 127, 136, 146, 149, 205
 disordered multistate, 34, 41–42, 75, 203
 extrinsic, 119, 126, 155, 156, 157
 graded, 35, 42–44
 intrinsic, 119, 126, 132
 meristic, 35, 41, 43, 45, 46, 48, 63, 87, 105, 127, 136, 139, 146, 148, 157
 missing, 36, 41, 75, 157
 mixed, 46–48, 75, 157
 multistate, 19, 34, 37, 46, 48, 75, 157, 203
 nonconsequential, 36
 non-exclusive multistate, 35
 number of, 33
 ordered multistate, 34, 42–44, 75, 204
 ordinal, 35, 42–44
 presence-absence, 13, 15, *see also* Binary
 quantitive, 125
 ranked, 35, 44–45
 single, 32, 33
 weighting of, 13
Avocado, 64

B

Bacteria, 118
Bees, 47, 64
Beetle, 36
Benthos (ic), 19, 22, 24, 28, 39, 40, 41, 47, 48, 62, 64, 85, 88, 89, 100, 108, 110, 112, 114, 116, 119, 123, 127, 132, 134, 143, 146, 149, 150, 155, 158, 164, 165
Binomial, 3–4
Biocoenosis, 17, 18, 20, 22, 149
Biogeography, 15, 16, 38, 108, 165
Biomass, 39

Biome, 19
Biota, 17, 143
Biotope (ic), 17, 18, 20
Birds, 16, 100, 186
Braun-Blanquet, 17, 21, 23, 98
Brillouin, 68, 72, 85, 161, 168, 194
Butterflies, 163

C

Canonical correlation analysis, 169, 184
Canonical variate analysis, 169, 183
Charybdis, 45
Cell density (in two-way table), 132, 133, 135
Citrus, 3, 4
Classification(s)
  agglomerative, 4, 13, 29, 100, 139
  association analysis, 103, 155
  biocoenotic, 20–24
  biogeographical, 15
  comparison of, 125
  concepts of, 26
  divisive, 20, 29, 88, 100, 117, 139, 155, 157
  ecological, 20, 45
  environmental, 21
  exclusive, 27
  extrinsic, 27, 157
  general, 1, 26–27, 158
  hierarchical, 28, 104, 138
  history of, 3
  intrinsic, 27
  inverse, 16, 19, 34, 87, 134
  monothetic, 13, 20, 29, 88, 100, 101, 155, 157
  non-exclusive, 27
  nonhierarchical, 28, 133
  normal, 16, 19, 34, 87, 98, 136
  numerical, 31, 33
  objectives in, 143
  polythetic, 13, 29, 100, 104, 117
  "Q," 34, 127
  "R," 34, 127
  synecological, 21
  taxonomic, 48
  use of names, 17–20
Cline, 119
Clumping, 30, 41, 118, 119, 127
Clustering strategies, *see also* Classification
  centroid, 110, 116
  chaining, 106
  combinatorial, 105, 107
  comparison of, 139
  definition, 30
  density, 117
  flexible, 114, 125, 133, 137, 150
  furthest-neighbor, 109, 139, 140, 142
  general, 30
  group average, 113, 125, 133, 137, 139, 140, 142
  incremental sum of squares, 113, 116
  mean sum of squares, 114
  median, 112
  minimal variance, 114
  monothetic divisive hierarchical clustering methods, 101–104
    based on
      χ^2, 103–104
      information measures, 102–103
  nearest-neighbor, 107, 121, 124, 139, 140, 142, 147
  probabilistic, 118
  single linkage, 107, 108
  space conserving (dilating and contracting), 106
Community, 16, 17, 21, 143, 144, 149, 163, 164, 165, 167
Conifers, 104
Connectedness, 121
Constancy, 12, 22, 23, 39, 51, 90, 100, 101, 103, 127, 132, 135, 138, 147, 148
Contiguity, 1, 2
Continuum (a) (continuous systems), 2, 10, 11, 24, 25, 42, 45, 189
Cophenetic correlation coefficient, 142
Corals, 40, 44, 89, 149
Crabs, 15, 38, 45

D

Data, *see also* Attributes
  biographical, 40
  centering, 96, 147, 150, 175, 183, 187
  conversion, 46
  ecological, 40
  extrinsic, 126
  general, 32–34

intrinsic, 126
mixed, 46, 47, 102
raw, 43
reduction, 7, 83, 85, 119, 125, 137, 147, 150, 190
scaling, 85
standardization, 45, 83, 93, 125, 145, 147, 153, 173, 183, 187, 188
storage–recovery, 13
transformation, 43, 83, 89, 90, 125, 146, 147, 150, 153, 158, 159, 163, 170, 187, 188
Decision function, 76
Dendrogram, 9, 105, 107, 120, 124, 133, 134, 137, 138, 139, 140, 141, 142, 169, 187, 190
Dicotyledons, 104
Discrimination, 30
Disorder, 68
Dissection, 26, 27, 30
Dissimilarity, 50, 124, *see also* Similarity
Dissimilarity measures, *see also* Similarity measures
Bray-Curtis, 57, 58, 80, 84, 90, 95, 125, 138, 153, 155, 163, 182, 188, 189
Cain and Harrison, 60
Canberra metric, 58, 80, 85, 86, 90, 91, 95, 138, 150, 151, 152, 163, 182, 189
coefficient of divergence, 60
between dendrograms, 139
Euclidean distance (D, D^2), *see* Euclidean distance
Gower, 61
information theory, 67
Klauber, 60
mean character difference, 60
Distance, *see also* Euclidean distance
ecological, 93
taxonomic, 9
Diversity
alpha (α), 167
beta (β), 167
Brillouin, 68, 72, 85, 161, 168, 194
dominance (-cy), 163, 167
gamma (γ), 167
general, 8, 42, 51, 68, 72, 90, 157, 160, 164, 165, 191, 196
partitioning, 167, 168, 196
per individual, 68
per site, 68
Shannon, 68, 71, 72, 85, 161, 192
species, 163, 167
values, 158
Dominance, 23, 39, 44, 45, 90, 127, 135, 138, 148, 160, 162, 163, 164, 167

E

Earthworms, 114
Ecology, 38
diffuse nature of, 16–17
Ecosystem, 16, 17, 18, 164
Edden, 160
Eigen value, 175
Eigen vector, 174, 175
Elytra, 35, 36
Entity, 2, 10, 20, 26, 27, 33, 90, 127
Equitability (evenness), 161
Erianthus, 119
Estuary (ine), 18, 19
Eucalyptus, 97, 178, 183
Euclidean distance (D^2), 49, 50, 51, 64, 65, 78, 79, 80, 85, 89, 90, 93, 94, 98, 113, 139, 142, 150, 153, 162, 163

F

"F" test, 136, 137
Factor analysis, 170, 178
Faithful, 12
Felspar, 25
Fidelity, 12, 23, 39, 90, 132, 135, 138, 148, 163
Field mouse, 183
Fisher, 159
Fission, 4
Formation, 19, 22
Fossils, 166

G

Gazelles, 183
Genealogy, 9
Genotype, 97
Geomorphology(ical), 18
Graph theory, 120, 138
minimally connected, 122
Grasses, 103, 104, 108, 112, 118, 123
Group-size dependence(-cy), 105, 133, 137

H

Halophila, 48
Hierarchy, 7
Homogeneity, 121
Hurlbert, 160, 161

I

Identification, 26, 27, 30
Indicator species, 40, 155
Information, *see also* Diversity
  content, 51, 100, 139
  gain, 70, 71, 79, 104, 105, 125
  measures, 67, 102
  partitioning, 72
  statistic, 47
  theory, 16, 42, 136, 191

K

Kruskal-Wallis, 136, 137, 157

L

Latent roots, 175
Latent vector, 174, 175
Legumes, 104
Lepidoptera, 16
Linnaeus, 3–6, 21
Lomandra, 114

M

McIntosh, 160, 162, 163
Malthus, 21
Mango, 64
Mangrove, 40
Margalef, 159
Matrix, 33, 39, 99, 126, 157, 180, 181, 182, 197
Metric, 78, 107, 208
Misclassification, 127
Models
  1-dimensional, 8
  2-dimensional, 2, 8–9, 169, 187, 190
  3-dimensional, 9, 187
Monothetic, *see* Classification
Monotonic, 51, 52, 107, 111

N

Nekton, 39
Nematodes, 88
Niche, 24, 166
Nomenclature, 3
Normalization, 96

O

Objectivity, *see* Subjectivity
Orchids, 104, 113, 114, 123, 187
Ordination, 27, 30, 124, 133, 134, 135, 169
  Bray and Curtis, 177
  interpretation of, 136
Ostracods, 88
O.T.U., 20

P

Paleontology, 9, 11, 182
Paleospecies, 11
Paleotaxonomy, 25
Pattern, 38
Phyllota, 117, 118
Phylogeny, 7, 10, 11
Phytosociology, 22, 46, 147
Pielou, 160, 162
Planets, 170
Plankton, 22, 40, 99, 100, 108, 165
Polynomial, 4, 6
Polythetic, *see* Classification
Principal component analysis, 169
Principal coordinate analysis, 169, 182
Probability (probabilistic), 51, 77, 118, 160
  estimates, 51
Psychology, 25, 31, 97

Q

Quadrat, 45

R

Rain forest, 41, 46, 119, 186
Rarefaction, 158, 159
Reallocation, 100, 103, 117, 127, 137, 138
Redundancy, 162
Regression, 171, 172
Reversal, 111
Rhizobium, 108, 110

S

- *Saccharum*, 119
- *Salvia*, 112, 116
- Sample, 45
- Sampling, 144
- Sedges, 104
- Shannon, 65, 68, 71, 72, 161, 192
- Similarity, 36, 49, 50
- Similarity measures, *see also* dissimilarity measures
  - application to
    - binary data, 52–57
      - with conjoint absences, 53–54
      - without conjoint absences, 54–57
    - continuous data, 58–61
    - meristic data, 58–61
  - coefficients of associations, 50
  - Czekanouski, 55, 79, 81, 87, 97
  - Fager and McGowan, 55, 79, 97
  - Hamann, 54
  - Hawksworth *et al.*, 71
  - information theory, 67, 207
  - Jaccard, 54, 79, 81, 87, 97
  - Kulczinski, 55, 97
  - Ochiai, 55, 97
  - Preston, 57
  - properties of, 77
  - simple matching, 53, 80
  - Sokal and Sneath, 54
  - Sørensen, 55
  - Rogers and Tanimoto, 54
  - Russell and Rao, 54
- Simpson, 160
- Snake, 60
- Sociology(ical), 25, 27
- Space
  - conserving, 106, 137
  - contracting, 106
  - dilating, 106
  - multidimensional (hyperspace), 9, 27, 66, 82, 117, 124, 135, 169, 184, 187, 188
- Species
  - biospecies, 11
  - codominant, 90
  - chronospecies, 11
  - indicator, 40, 155
  - nonseasonal, 149
  - paleospecies, 11
- Stability, 166
- Standardization, *see* Data
- Station, 45
- Statistics (statistical), 21, 31, 32
- Subjective (subjectivity), 6, 31, 43, 91, 125, 134, 137, 138
  - Succession, 22
- Systematics definition, 2

T

- Table
  - coincidence, 128–131, 134
  - two-way, 126, 127, 134, 152, 153
- Taxonomy
  - classic, 12
  - definition, 2
  - neontological, 11
  - numerical, 13
- Taxonomic
  - continuum, 10–11
  - distance, 9
- Termites, 112
- Tests
  - of significance, 136, 190
  - non parametric, 137, 157
- *Thalamita*, 45
- Theophrastus, 21
- Time factor, 11, 17, 20, 22, 23, 144, 146, 149, 157, 165
- Traverse, 144
- Trees
  - minimum spanning, 120, 123, 169
  - and non-trees, 1, 23
  - trellis diagram, 93, 94, 99, 124, 141
- *Trifolium*, 108, 110, 112
- Type specimen, 7

U

- Ubiquity(ous), 39, 40, 87, 88, 132, 137
- Unevenness, 162
- Uniformity index, 162

V

- *Venus*, 46

W

- Weighting
  - equal, 13
  - independent, 147
  - objective, 14